Management in the Built Environment

The aim of this book series is to provide a platform to build and consolidate a rigorous and significant repository of academic, practice and research publications that contribute to further knowledge relating to management in the built environment. Its objectives are to:

1. Disseminate new and contemporary knowledge relating to research and practice in the built environment
2. Promote synergy across different research and practice domains in the built environment and
3. Advance cutting-edge research and best practice in the built environment

The scope of this book series is not limited to "management" issues per se because this then begs the question of what exactly are we managing in the built environment. While the primary focus is on management issues in the building and construction industry, its scope has been extended upstream to the design management phase and downstream to the post-occupancy facilities management phase. Management in the built environment also involves other closely allied disciplines in the areas of economics, environment, legal and technology. Hence, the starting point of this book series lies with project management, extends into construction and ends with facilities management. In between this spectrum, there are also other management-related issues that are allied with or relevant to the built environment. These can include, for example cost management, disaster management, contract management and management of technology.

This book series serves to engage and encourage the generation of new knowledge in these areas and to offer a publishing platform within which different strands of management in the built environment can be positioned to promote synergistic collaboration at their interfaces. This book series also provides a platform for other authors to benchmark their thoughts to identify innovative ideas that they can further build on to further advance cutting-edge research and best practice in the built environment.

If you are interested in submitting a proposal for this series, please kindly contact the Series Editor or the Publishing Editor at Springer:

Low Sui Pheng (bdglowsp@nus.edu.sg) or
Ramesh Premnath (Ramesh.premnath@springer.com)

More information about this series at http://www.springer.com/series/15765

Syamilah Yacob · Azlan Shah Ali ·
Cheong Peng Au-Yong

Managing Building Deterioration

Prediction Model for Public Schools in Developing Countries

Springer

Syamilah Yacob
Public Work Department (PWD)
Cawangan Senggara Fasiliti Bangunan
Kuala Lumpur, Malaysia

Cheong Peng Au-Yong
Department of Building Surveying
Faculty of Built Environment
University of Malaya
Kuala Lumpur, Malaysia

Azlan Shah Ali
Department of Building Surveying
Faculty of Built Environment
University of Malaya
Kuala Lumpur, Malaysia

ISSN 2522-0047 ISSN 2522-0055 (electronic)
Management in the Built Environment
ISBN 978-981-16-5859-4 ISBN 978-981-16-5860-0 (eBook)
https://doi.org/10.1007/978-981-16-5860-0

This Springer imprint is published by the registered company Springer Nature Singapore Pte Ltd.
The registered company address is: 152 Beach Road, #21-01/04 Gateway East, Singapore 189721, Singapore

Summary

The inefficiencies of the building maintenance system could result in defects and damages to the building. Generally, most of the government buildings are not given extensive attention while maintenance work is carried out especially to school buildings. This matter has caused the cost of maintenance to increase dramatically as well as the expenditure that was invested in preparing the building. The problem can cause various effects and negative impact to the user and owner of the building if the problems are not addressed systematically. Thus, there is a necessity to enhance the current processes of maintenance system to improve the government service delivery. The triangulation approach was used in the study with a mixed of quantitative and qualitative methods for collection of data. Three hundred and three (303) school buildings were selected as case study based on population research. The schools were selected based on criteria set-up in the study such as the school age, types, level and region. Literature review and analysis of archived documentation were carried out to identify the factors that contributed to building defects. From the archived documentation also, the condition of the building was assessed using Public Work Department (PWD) Building Condition Assessment Rating Systems (BCARS) for data collection. Correlation analysis was then done to test the relationship between the variables identified and condition of the building using the statistical analysis software Statistical Package for Social Science (SPSS). In addition, the study also produced a regression model that can be used to forecast changes to the deterioration of the condition of the building.

The study identified twenty-one (21) reliable variables related to the factors that cause the building defects. The result analysis shows that the six (6) variables have a strong relationship with the building condition. The factors which have a strong significant correlation to building condition are lack of maintenance, vandalism, poor waterproofing, lack of supervision, lack of cleaning and misuse by users. The results also show that these factors are the main causes at most school buildings in Malaysia. These factors need to be taken into consideration in order to address the existing maintenance. Subsequently, the results produced the logistic regression model for three (3) variables that caused building defects towards building condition based on the following circumstances with the equation: $Z = -3.479 + 0.658\ (PW) + 0.755\ (LOM) + 0.445\ (LOC)$. In conclusion, a regression model produced will be a tool used to forecast changes or deterioration of the building condition in order to improve the current government maintenance system besides to make decisions that are more efficient especially for maintenance budgeting, repair work, and forecast standard lifetime building element can be specified.

Contents

Abbreviations

ASCE	American Society of Civil Engineers
BCA	Building Condition Assessment
BCARS	Building Condition Assessment Rating System
BCMAS	Building Condition Maintenance Assessment System
BDPM	Building Deterioration Prediction Model
CF	Certificate of Fitness
CFA	Confirmatory Factor Analysis
EFA	Exploratory Factor Analysis
GAMP	Government Asset Management Policy
GDP	The Convergence Model
GNP	Gross National Product
IIAM	Institute Infrastructure Asset Management
IIMM	Institute Infrastructure Management Manual
IPWEA	Institute of Public Works Engineering Australasia
KMO	Kaiser–Meyer–Olkin
MOE	Ministry of Education
MOHE	Ministry of Higher Education
MPAM	Manual Pengurusan Aset Menyeluruh
MySPATA	Sistem Pengurusan Aset Tak Alih
NAPIC	National Property Information Centre
NCES	National Centre Education System
PWD	Public Work Department
RICS	Royal Institute Chartered Surveyor
RMK	Rancangan Malaysia
SPSS	Statistical Package for Social Science
TAM	Total Asset Management
TAMM	Total Asset Management Manual
TPATA	Tatacara Pengurusan Aset Tak Alih
WHO	World Health Organization

List of Figures

List of Tables

Chapter 1
Introduction

Abstract This chapter explains the background of the study and highlighted the issue and rationale of the study. The research objectives were formulated from the research background, problem statement, rationale of the study and the area to improve the study. The benefit, the significance of the study, scope and limitation also indicates in this chapter to ensure the objectives of the research conducted accordingly hence to predict the condition of Government building. The chapter also focuses on the literature overview on the building defects which related to the study.

1.1 Background

The government built on the stock of infrastructure since expanded manifold and facilities have been modernised [1]. Infrastructure includes the broad categories of assets which provided the facilities, services and installations needed for the functioning of a community or society [2]. According to Institute Infrastructure Asset Management [3], infrastructure assets are motionless systems that present the communities and needed to be maintained to a specified level of service such as transportation, energy supply systems, parks and recreation facilities, water utilities, flood protection and land drainage systems, solid waste facilities, learning and health sector facility, libraries, administration and other community facilities, manufacturing and telecommunications. Infrastructure assets include both physical and non-physical assets such as building and software and programs. Meanwhile, Total Asset Management Manual [4], defined infrastructure asset as network systems that serve the communities such as road, bridge, airport, jetty, energy and water supply, telecommunication, drainage, sewerage system and engineering assets such as building, plant, and others.

Malaysia is a developing country and continues to grow to have an important influence on infrastructure or asset development in the country for five decades from 1966 to 2016. The three aspects which have an important influence such as growth performance, physical make-up and socioeconomic disparities which must take into

S. Yacob et al., *Managing Building Deterioration*, Management in the Built Environment,
https://doi.org/10.1007/978-981-16-5860-0_1

consideration in the formulation of infrastructure policies [1]. The Malaysian Government built on the stock of infrastructure or asset, expanding and modernising infrastructure facilities [1] to secure the interest of citizens in terms of providing quality services to the people through the provision of proper infrastructure [5]. Mohd Isa et al. [6] states by the year 2017, the government has committed with the allocation of RM 1079 million for upgrading and renovation. Based on the Financial Report of Ministry of Finance for the year 2010 until 2015, the government has spent the sum of RM 292,670 million for public sector development expenditure which at the current exchange rate is equal to USD 65,687 million and spent the sum of RM 61,004 million for upgrading, renovating and maintenance of the various facilities. The renovating and maintenance budget have kept rising from 12% in 2010 to 39% in 2015 as part of the development budget.

School buildings in Malaysia have been developed and grown as national assets that the government is committed to ensure the safety and the maintainability of the buildings. Schools form the largest portion of non-residential building registered in the Directory of Federal Government Premises Registration (JKR 21600–0012-12) with 10,381 in total. The importance of education sector is signified by the amount of budget allocation by government [7]. The government has allocated more than 20% of the national budget for education and training and in an average of 5% of Gross National Product (GNP). Malaysian Plan developed is to improve the quality of students which include to create a conductive learning environment that will affect students' performance. Over the years, students' enrolment in public education has increased during the 2000–2005 period by 9.9% and expanded 8.8% between 2005 and 2010 [8].

Under the 10th Malaysia Plan, the number of high-performing schools increased to 100 by the end of 2012 which include primary, secondary, day and residential schools. Under the 11th Malaysia Plan, the government has made large investments in enhancing the people's wellbeing under construction project development programs. One of the projects has been planned for the next 5 years to improve the people's well-being is to build 80 new primary and secondary schools. Besides that, the government also made large allocations to school building under Malaysia Economic Stimulus Package which amounted to RM 565 million for maintenance since 2009. The asset required acceptable service level to ensure the functionality of the assets. Mohseni [9] reported the asset required service level with minimum cost and an optimized financial plan for whole life cycle of the assets. Meanwhile Institute of Public Works Engineering Australasia [3] reported that infrastructure assets required to be operated, maintained, replaced, enhanced or disposed to provide the required service level.

1.2 Problem Statement

The practice of asset management in Malaysia focuses on the building operation, maintenance management, space management, and security management and was lacking on performance monitoring [10]. The practice and performance of government assets and public assets typically adopted a reactive maintenance, ad hoc without systematic plan and schedule [5]. The reactive maintenance practice caused lower asset life, incur high maintenance cost, reducing the quality of services and others.

There are numerous defects most arise at school buildings and being reported officially by mass media [11]. The school buildings in Malaysia were not in good condition and reported most schools do not have their own maintenance plan but depended more on emergency planning [12]. The statement by Che Ani et al. [5] indicated the school as one of government biggest asset and should be maintained to ensure its functionality and value. According to Yacob [13], in Malaysia, school building maintenance are usually neglected and in unsatisfactory condition. Previous research also has shown that the highest number of building owned by government in Malaysia was school building and its audited and reported in poor condition, hence the school maintenance still practiced in an unsatisfactory level, which lead to the increment of the maintenance cost of school buildings [14].

Systematic approach towards building management is crucial [9]. The decision of asset managers is becoming more complicated and intense knowledge of the asset condition is needed [14]. The effective maintenance work depends on accurate data and information about the current building condition [15]. The maintenance data and information about the building obtained by building inspection and the building condition should be maintained to ensure its functionality and maintain the value [5]. In response to these challenges, asset management systems including assessment of current condition; a future deterioration prediction of building; and the choice of maintenance and strategies have evolved [2].

In this study, school problems regarding its condition were studied and the problem lies in the building defects. A lot of contributing factors towards the building defects were during the construction and pre-occupancy stage. Furthermore, discussions between factors contributing towards the building defects and building condition using building condition assessment (BCA) tools for existing building were discussed. The BCA tools will be discussed especially in its application for buildings in Malaysia. Samah et al. [16], has conducted a study using condition survey protocol matrix to determine the condition of school buildings to prove that the age of the school has a significant relationship with its' condition. However, there have yet to be any building conditions conducted using BCA JKR 21602-0004-13 especially for public buildings in Malaysia.

There are also some related works regarding the building prediction model. According to Mohseni [9], the deterioration of component is predicted using discrete condition data collected through visual inspection. Other study on the prediction model in practice carried by Koru and Liu [17] focuses more to application software. Turhan et al. [18] reported that, defect prediction is still open challenge and should be considered in other situations and environments. Prediction model is also used to study concrete bridges using factor method based on historic inspection data [19]. Meanwhile, Mashford et al. [20] focuses more on sewer asset, and Basheer et al. [21] focused more on the concrete structure prediction model.

1.3 Research Questions

To begin the study, certain questions should be addressed before the objectives and conclusion can be finalized. The following questions were identified as follows:

(i) What are the factors affecting school building deterioration in Malaysia?
(ii) What is the current state or level of conditions of school buildings in Malaysia?
(iii) To what extent is the relationship between the factors and building condition?
(iv) What is the factors method used for prediction based on building condition?

1.4 Research Aims and Objectives

The research aims to develop a building deterioration prediction model (BDPM) of the building conditions based on factors contributing towards building defects for school buildings.

The objectives were formulated as below:

(a) To identify the factors contributing towards building defects;
(b) To investigate the condition of public-school buildings through building condition assessment;
(c) To establish critical factors contributing towards school buildings defects and its' relationship to building condition;
(d) To develop a building deterioration prediction model (BDPM) of the building condition.

1.5 Rationale of the Study

The past decade has discussed about the defect factors and the need for more attention to address and to overcome the issues. The recurrent incidence of defects in a building especially for schools contributed to high maintenance cost for school building in Malaysia [22]. There are various reasons for defects to occur and has negative impact to school buildings [23]. As such, it is crucial to investigate the factors contributing to the defects to create a safe building [23–25] states that a contributing factors to building failures will improve the effectiveness of managing the appraisal process of failure and defects in the future.

Defects occur in building for many reasons. This is due to various ranges of factors contributing to the building defects. The defects in building occur because lack of maintenance, lack of supervision, material, faulty in construction, faulty design, building types and location [25]. Meanwhile Chong and Low [26] reported that the factors contributing to the building defects are poor quality of material, weather, wear and tear, impact from the occupant, vandalism and moisture. Another study conducted by Kishore Kumar and Sriram [27], found that major defects in building occur due to user mistakes, poor workmanship, and the uses of substandard materials. Meanwhile Othuman Mydin et al. [28] found that the main factor in the presence of building defect is faulty design.

From the literature, it is found that lack of maintenance was most reported as factor contributed to building defects. As reported by Shanker [29], Malaysians have lack of maintenance culture. This issue is also highlighted by Lam [30], poor maintenance of public building and infrastructure often lead to accidents and injuries. Lack of maintenance was ranked as the most significant factor contributed to building defects [31]. Thus, from the literature, it showed that the contributing factors were the main reason defects to occur. The emergence of defects affecting the condition of the building and the comfort of occupants. The summary of the literature review above has proven that it is essential to identify the factors contributing to the building defects in order to minimize the effect to building especially for school building. The literature search also showed that a systematic approach to building management, such as maintenance, is necessary to extend the life span of the building. Therefore, the rationale for conducting this study is to improve the government service delivery system toward effective decision making for building maintenance.

1.6 The Area for Development

Building condition assessment (BCA) has advantages for the management and maintenance of the building. She also mentioned that BCA is based on qualifications and quantification of defects in building and system component and it reveals maintenance activities that must be carried out. BCA was developed by PWD Malaysia to improve the government service delivery toward effective decision making for building maintenance. From the literature review, BCA determined by inspection and the causes, the defects and their characteristic is identified and listed. However, it still needs to review in order to enhance government maintenance processes. From a preliminary study by Public Work Department (PWD), there is no reliable source of information about contributing factors to building defects provided and kept by the state government for forecasting building deterioration especially for school building.

To enhance the quality of maintenance management processes, the development of the BDPM using building deterioration curves is vital to improve the current practices of maintenance management for buildings [9]. According to Keshavarzrad et al. [32], deterioration prediction as a forecasting tool, possible to track the condition of the assets without the need of initial physical inspections and when no reliable condition data are available. Sujova et al. [33], reported that the process improvement is to achieve the change, increasing process level and performance which increase the value added to the organization. Meanwhile Edirisinghe et al. [34] noted that prediction model is to predict the deterioration of the building which gives a step change in building management. Prediction is essential to improve proactive maintenance [35].

The prediction model indeed can improve organizational decision making. Therefore, it is necessary to study the factors contributing to building defects towards building condition to predict the condition for better maintenance.

The study conducted is to improve the current processes of maintenance management, decision making and increase the performance, which to improve the government service delivery. For this study, by using the scientific method in predicting the building deterioration, the following areas will be improved:

(i) The maintenance management performance [15];
(ii) The current process of building condition assessment [1];
(iii) The organization decision making;
(iv) Reduction in additional budget and saving in time, resources, operation budget and maintenance cost [15];
(v) Performance of the organization and the management [33].

1.7 The Significant of the Study

The finding of the study can contribute greatly to the purpose of the organizations which is the BDPM can plays an important role in managing the asset especially for building maintenance. Development of the BDPM in controlling the maintenance cost, time, accuracy, resources, performance, and service life prediction of the building [15]. This will help the organizations to monitor the condition of the building and predict the maintenance expenditure.

This research not only contribute to knowledge and theory, but it is also contributing to good maintenance practices and derive to decision making strategy for the organizations [33]. The outcomes of the research will benefit the organizations as followed:

(i) The study will seek to find out the various factors that caused the defects, and this will contribute to appropriate remedial action to be taken to overcome the building deterioration and defects;
(ii) Identification of the defects and factors should be able to help clients and designers to come out with appropriate and systematic strategies to improve maintenance;
(iii) BDPM is for a decision-making strategy to the organizations and buildings and also provide guides of the best practices in asset management;
(iv) Assisted to become aware the condition of buildings which can effect on the safety and health of the person, prevent the buildings which can lead to the increment of the maintenance costs;
(v) Provide critical decision making to carry out systematically, effective and efficient maintenance of the buildings;
(vi) To improve maintenance guideline by developing a scientific method in predicting building deterioration.

1.8 Limitation and Scope of Study

In this study, the limitation and scope of selecting school buildings are as follow:

(i) The focus of this research is to study the factors contributing to building defects, it has been limited to primary and secondary school buildings; National School and National-type School only, excluding other types of school such as Cluster, Vision, Islamic Religion, Bestari, Vocational, Technic and Boarding Schools. The National and National–type school was chosen due to the standard design produced by PWD. It is important to study the factors contributing towards building defects for schools designed by PWD for future improvement, especially as deliverables of maintenance for future design.

(ii) The selection of schools was based on PWD monitoring especially for maintenance work. The government's general circular letter No. 2 year 1995 (Surat Pekeliling Am Bil.2 Tahun 1995) instructed all government agencies to give greater emphasis to strengthen the maintenance management system in the public service. This includes the appointment of contractors, budget preparation and allocation monitoring. MOE needs to request the services of PWD as the technical agency to advice, monitor and supervise in related fields building maintenance. The responsibility to monitor is performed by the PWD of the state or district. Hence, the supervision work by PWD is very important to ensure that school buildings are well maintained and are in good condition.

(iii) The inspection carried out by PWD in 2010 using Building Condition Maintenance Assessment System (BCMAS) is to ascertain the conditions of the relevant federal government buildings including schools. The study involved in analysing the secondary data kept by PWD is crucial to ensure that the data used in the study are reliable and can be used to improve the government service delivery. The data was collected and obtained from archived document kept by the PWD and no fieldwork was involved in this study. The criteria selection of schools was based on school age of 6 to 10 years after the inspections were carried out. The study also excluded newly completed schools because the older the school is, the more could have defects occurred [12]. Commonly, the critical school buildings are aged between 6 and 20 years [12]. Thus, the reliability of the data depends on the limited sources of archived documents used in this study.

(iv) Hussin et al. [36], stated in his research that the responsibility to maintain the school buildings in Malaysia is under MOE. The school buildings were the largest numbers built in more than 45 years and in need of extra care [1]. The assessment of the schools in Peninsular Malaysia is essential to ensure the data collected coverage the state in Malaysia, excluding Sabah and Sarawak since there are no inspections reported using BCMAS available for Sabah and Sarawak kept by PWD.

(v) The assessment of the buildings focused on the available governments' current method of inspection which covered the guidelines, procedures, reports and others. This study will not discuss other methods which are not enforced and used for government buildings.

1.9 Organization of Chapters

This book is structured into 7 chapters.

In this chapter; describes the background of the study, followed by problem statement of the study. The research questions aim, and objectives also explained in this chapter. The rationale of the study, the area to improve of the research as well as the significance of the research, limitation, and methodology are discussed and addressed in the thesis.

Chapter 2; describes the overview about the building defects, building stocks and the maintenance of the building. The studies on the theoretical of the building deterioration or defects occur in the buildings and it involved the mechanism of defect formations and types of defects. In this chapter also covered the explanation of possible causes of building defects.

Chapter 3; provides the literature studies and analysis, which covered the aim of gathering information, knowledge based on the objectives studied. It also described the phenomenon defects which consist the various factors contribute to building deterioration toward the condition of the building. The selection of variables explained in this chapter.

Chapter 4 reviews the current practices of building condition assessment in Malaysia. Building Condition Assessment (BCA) used as a tool to identify the current building condition. It also describes the design and implementation of the methods and frameworks. In this chapter also covered about the building hierarchy and building condition assessment (BCA), explores the BCA techniques, rating systems, inspection, data collection, analysis and current practices of BCA.

Chapter 5; explains the research methodology used for the research. The chapter explained on the research design of the study, which consists of two types of research, quantitative and qualitative, explained through the triangulation design visual model. Field research on the current condition of the building resulting from various factors contributed to the building defects. The studies and use of the computer program for collecting the building condition and ratings.

Chapter 6; presents the analysis and research findings of the research by using Statistical Package for the Social Science (SPSS) software. Selected appropriate statistical tests were employed using the software, based on the types of data and approach of hypothesis testing. In this chapter also explains the result findings of descriptive result, correlation, logistic and regression result. The validation of the results through semi-structured interview also explains in this chapter.

Chapter 7; provides the summary of the objectives and general conclusion the findings of the research, synthesis the general conclusions from the research and finding, seeks to further research in the area and offer recommendations for future research.

References

1. Naidu, G., (2008). Infrastructure development in Malaysia. In N. Kumar (Ed.), *International infrastructure development in East Asia—towards balanced reional development and integration*. ERIA Researh Paper Project Report 2007-2, Chiba IDE-JETRO, pp. 204–227.
2. Ahluwalia, S. S. (2008). *A framework for efficient condition assessment of the building infrastructure*. University of Waterloo.
3. INGENIUM. (2006). *International infrastructure management manual*. Public Work Engineering of Australia.
4. PWD. (2009). *Total asset management*. Malaysia Public Work Department.
5. Che Ani, A. I., Sairi, A., Johar, S., Razak, M. Z., & Tawil, N. (2015). The practice of building inspection in malaysia total asset management. *Jurnal Teknologi*, 1–11.
6. Mohd Isa, H., Hassan, P., Mat, M. C., Isnin Z., & Sapeciay, Z. (2011). Tracking architectural defects in malaysian hospital projects. In *IEEE Symposium on Bussiness, Engineering and Industrial Applications (ISBEIA) Langkawi* (pp. 298–302).
7. Mazlan, R. R., Mohammed, A., & A, B. (2008). Public school maintenance: Competency issue. In *2nd Engineering Conference on Sustainable Engineering Infrastructure Development & Management*. ENCON2008-244
8. Olanrewaju, A., & Aziz, A. A. (2015). *Building maintenance processes and practices*. Springer Science and Business. https://doi.org/10.1007/978-981-287-263-0-4
9. Mohseni, H. (2012). *Deterioration prediction of community building in Australia*. PhD unpublished Thesis, RMIT University.
10. Yusof, Z. A., & Bhattasali, D. (2008). *Economic growth and development in Malaysia: Policy making and leadership*. The World Bank.
11. Bakri, N. N. O., & Mydin, M. O. A. (2014). General building defects: Causes, symptoms and remedial work. *European Journal of Technology and Design, 3*(1), 4–17.
12. Mahli, M., Che-Ani, A. I., Abd-Razak, M. Z., Tawil, N. M., & Yahaya, H. (2012). School age and building defects: Analysis using condition survey protocol (CSP) 1 matrix. *International Journal of Civil, Environmental, Structural, Construction and Architectural Engineering, 6*(7), 56–58.
13. Yacob, S. (2005). *Maintenance management system through strategic planning for public school in Malaysia*. UTM.
14. Dejaco, M. C., Cecconi, F. R., & Maltese, S. (2016). Key performance indicators for building condition assessment. *Journal of Building Engineering*.
15. Yong, C. Y., & Sulieman, M. Z. (2015). Assessment of building maintenance management practice and occupant satisfaction of school buildings in Perak, Malaysia. *Jurnal Teknologi*, 1–6.
16. Samah, A. H., Tawil, N. M., Mahli, M., Che-Ani, A. I., & Abd-Razak, M. Z. (2014). Building condition assessment using condition survey protocol matrix: A case of school building. *Research Journal of Applied Science, 9*(9), 565–572.
17. Koru, A. G., & Liu, H. (2005). Building effective defect-prediction models in practice. *IEEE Software, 22*(6), 23–29.
18. Turhan, B., Tosun, A., & Bener, A. (2011, August). Empirical evaluation of mixed-project defect prediction models. In *2011 37th EUROMICRO Conference on Software Engineering and Advanced Applications* (pp. 396–403). IEEE.

19. Abu Backer, M. M. A., & Wan Yusoff, W. Z. (2014). A review of asset management in Malaysia local government. In *Proceedings of the 25th International Business Information Management Association Conference—Innovation Vision 2020: From Regional Development Sustainability to Global Economic Growth, IBIMA 2015 2015* (pp. 3577–3582).
20. Mashford, J., Marlow, D., Tran, H., & May, R. (2010). Prediction of sewer condition grade using support vector machines. *Journal of Computing in Civil Engineering*, 25. https://doi.org/10.1061/(ASCE)CP.1943-5487.0000089
21. Basheer, P. A. M., Chidiact, S. E., & Long, A. E. (1996). Predictive models for deterioration of concrete structures. *Construction and Building Materials, 10*(1), 27–37. ISSN 0950-0618. https://doi.org/10.1016/0950-0618(95)00092-5.
22. Ali, A. S., Keong K. C., Zakaria, N., Zolkafli, U., & Akashah, F. (2013). The effect of design on maintenance for school buildings in Penang, Malaysia. *Journal Structural Survey*, 194–201.
23. Adedeji, I. K. B. (2016). Assessment of contribution factors to defects in hall of residence in two selected tertiary institution in South West, Nigeria. *IOSR Journal of Humanities and Sosial Science (IOSR-JHSS), 21*(8), 58–64.
24. Chong, W. K., & Low, S. P. (2005). Latent building defects: Causes and design strategies to prevent them. *Journal of Performance of Constructed Facilities*, 283–289.
25. Ahzahar, N., Karim, N., Hassan, S., & Eman, J. (2011). A study of contribution factors to building failures and defects in construction industry. procedia engineering. In *The 2nd International Building Control Conference* (Vol. 20, pp. 249–255). Elsevier. https://doi.org/10.1016/j.proeng.2011.11.162
26. Chong, W. K., & Low, S. P. (2005). Assessment of defects at construction and occupancy stages. *Journal of Performance of Constructed Facilities*, 283–289.
27. Kishore Kumar, S., & Sriram, P. (2017). Pivotal factors causing defects in buildings: A case study in Chennai, Tamilnadu. *International Journal of Civil Engineering and Technology, 8*(4), 325–331.
28. Othuman Mydin, M. A., Agus Salim, N. A., Tan, S. W., Tawil, N. M., & Md Ulang, N. (2014). Assessment of significant causes to school building defects. In *Emerging Technology for Sustainable Development Congress (ETSDC)*, EDP Sciences, (pp. 1–7).
29. Shanker, S. (2017). *Malaysian maintenance culture, or lack of it*, NST July 13 2017.
30. Lam, E. S. (2007), *Water seepage, flood and hydraulic*. Paper presented at Structure Congress, 16–19th May, 2007, Unites States.
31. Talib, R. (2014). Assessment of factors affecting building maintenance and defects of public buildings in Penang, Malaysia. *Journal Scientific Academic Publishing*, 48–53.
32. Keshavarzrad, P., Setunge, S., & Zhang, G. (2014). *Deterioration prediction of building components*. School of Civil, Environment & Chemical Engineering, RMIT University.
33. Sujova, A., Simanova, L., & Marcinekova, K. (2016). Sustainable process performance by application of six sigma concepts: The research study of two industrial cases. *Sustainability, 8*(3), 260. https://doi.org/10.3390/su8030260
34. Edirisinghe, R., Setunge, S., & Zhang, G. (2015). Markov model—based building deterioration prediction and ISO factor analysis for building management. *Journal of Management in Engineering, 31*(6), 04015009.
35. Abu-Samah, A., Shahzad, M., Zamai, E., & Said, A. B. (2015). Failure prediction methodology for improvement proactive maintenance using Bayesian approach. In *International Federation of Automatic Control, IFAC 2015* (pp. 844–851). Elsevier Ltd.https://doi.org/10.1016/j.ifacol.2015.09.632
36. Hussin, N. H. M., Sann, L. M., Shamsudin, M. N., & Hashim, Z. (2011). Characterization of bacteria and fungi bioaerosol in the indoor air of selected primary schools in Malaysia. *Indoor and Built Environment, 20*(6), 607–617.

Chapter 2
An Overview and Understanding the Building Deterioration

Abstract This chapter reviews the literature related to the introduction, overview on defects and maintenance management in general. The literature to building stock are discussed in this chapter to provide an understanding to the readers on the current scenarios of the general information on buildings in Malaysia. This overview is necessary to understand the knowledge of the related topic for in depth study. Then the study on the school buildings are explained in this topic to understand the basic concept of the education system in Malaysia. It covered the enrolment, age of students, types of schools, condition and performance of schools. This chapter also discusses the types of defects and these will be one of indicators which represent the variables in the research. Defects can arise due to many factors which will be explained in the next chapter. An understanding of the building defects and their causes is essential for better performance of any building and it is also a logical way of proceeding from the evidence to the cause of a defect, after which remedies can be prescribed. Appropriate solutions to mitigate the defect should be carried out to ensure the building are in a good performance and condition.

2.1 Introduction

In principle buildings should save and protect the occupant from various weather conditions, especially for the building element and services in the building. It should be serviceable to suit the facilities in the buildings, to ensure the building is safe and in a long-term quality. Factors that caused the defects should be prevented at pre and post occupancy stage. Building defects are common and can occur in various forms and to different extents in all types of buildings without considering the age and the construction of the buildings [1]. Inefficiency in handling the defects will generate several effects and negative impact to the user and the owner of the buildings.

According to Vitelli and Nelson [2] defects in buildings harm the reputation of the construction industry and the number of defects is believed to represent a loss in economy. Vitelli and Nelson [2] also defined defects as failure or malfunction with good ability, in terms of building guidelines or user needs in the building that may affect the structure, outline or fabrics, building services or facilities. Building defects

S. Yacob et al., *Managing Building Deterioration*, Management in the Built Environment,
https://doi.org/10.1007/978-981-16-5860-0_2

refers to the elements of the building, e.g., floor finishes and roof coverings. Building failure more like failure structure, for example the movement of soil, deposition of base buildings, design failure, for example leaks at the connection wall/floor. Meanwhile the malfunction of the building refers to services does not work; in particular to mechanical/electrical equipment. Deteriorate refers to degradation quality building materials in line with the age of the building.

2.2 Overview on Defects and Maintenance Management

The issues related to building maintenance in Malaysia were practiced and highlighted in the Second Malaysia Plan (1971–1975) but the issues did not receive adequate attention for almost 20 years [3]. The allocation for repair and maintenance increased from RM296 million in 8th Malaysian Plan to RM1, 079 million in the 9th Malaysian Plan. Rusli et al. [4] reported that repair and maintenance activity become increasing important in the future. Ahluwalia [5] cited the study by Kyle et al. [6], in order to preserve an asset in its original condition, the execution of maintenance activities such as inspection, preventive maintenance, repair, and rehabilitation should be carried out. The non-performing towards building maintenance work could result in defects as well as building failures. Both elements of defects and building failures as described as the building which failed to perform and delivered as it was planned. These problems happened to most of buildings without considering their age and type of construction. Inabilities in handling those building defects and building failures systematically provided negative impact to users and building owners as well.

Most of the government buildings need for the extra care and maintenance. Shittu et al. [7] mentioned that they are faced with lots of problems such as "cracking, staining/discoloration, sealant failures, efflorescence, rising dampness/water penetration, corrosion, buckling/deflection, tile/plaster delaminating" which have resulted due to poor building materials and workmanship. According to Hermawan [8] defects defined imperfection; blemish; fault due to lack of design, manufacturing process or improper installation of materials. Meanwhile Ahzahar et al. [9] also mentioned that building failure and defect influence by climate conditions, building location, material of construction, type of building and functional changes, building maintenance, design fault, corruption and lack of supervision.

According to Wong et al. [10], defects are a common phenomenon in Malaysian construction projects. He adds that most of rectification work relates to maintenance and these have caused the government a lot of money. Mohd Isa et al. [11] also mentioned that the Malaysian government has continually provided allocations for maintenance works for buildings and most of the money was spent to rectify defects. Many buildings in Malaysia are under or wrongly used and lack of maintenance [12, 13]. This would induce the building to become dilapidated. The lack of maintenance coupled with social problems, which can lead to vandalism and abuse increase the rate of wear and tear of buildings. Ali et al. [14] in his research, found that buildings have not been effectively maintained throughout the years, though there is a commitment from the government towards building maintenance.

Roslan et al. [15] asserted that maintenance problems in the building are common regardless of the size and the owners of the buildings. The importance of maintenance has been neglected by the building users. Building maintenance in Malaysia is conditionally driven and is usually concerned when there is cost allocation [16]. Based on his research also Olanrewaju [16], stated the government of Malaysia has committed and spent US$330 m for upgrading, renovating, and maintaining as part of development budget.

There is no proper planning on building maintenance [3]. Unplanned maintenance is often carried out instead of planned maintenance. The biggest problem is how and where to apply the available funds to ensure the maximum benefit. To overcome the problem of inaccurate maintenance information, the use of appropriate technologies and techniques for the information gathering such as building survey are required as mentioned by Minato [17]. Thus, it is important to have tools and techniques to support decision related to building maintenance and repair within the limited budgets of organisations.

Maintenance became more active in the year 2007 when a clear pathway for maintenance practices was highlighted in National Asset Facility & Facility Management (NAFAM) conference 2007. Hamimah [18] reported that maintenance management plays an important role to determine the success of maintenance work carried out by the organization. Her study also highlighted that the effectiveness of the maintenance is to have good management applied to the maintenance work. Meanwhile, Zawawi et al. [19], highlighted the study by Pintelon [20], which stated the excellent of the maintenance can be achieved by having the maintenance management that match with the requirements of the user.

The government has started making drastic changes in management by implementing the Total Asset Management (TAM) in all government buildings. The implementation of TAM is important for sustainable development [21]. The Total Asset Management Manual (TAMM) was introduced and implemented by the Government of Malaysia under the Government Asset Management Policy (GAMP) in 2009 to manage the government assets. The aims of TAMM are to realize the goal of GAMP by providing the quality services to the citizens through the provision of proper assets. It also emphasizes on the management of the government property assets in a systematic and holistic way to achieve optimum of the asset [22].

Currently, the practice of TAM in Malaysia gives each agency the responsibility to manage and maintain their assets. The effectiveness of asset management is to ensure the asset and facilities provided are manageable and can serve the purpose for which it was constructed. According to Ahluwalia [5], managing buildings are highly challenging due to the generally large size of the facilities, complex nature, and high costs. Asset management has evolved and its main functions include the assessment of current conditions; the selection of maintenance and repair strategies; the improvement of the condition after the repair; and the prioritisation of which building components should be repaired, given the budget constraints.

2.3 Building Stock in Malaysia

It is commonly recognized that existing building stock management has a great potential to achieve better sustainable development results. The construction of the building, whether by the government or private sector, is usually based on fulfilling specific needs, providing services, investment, occupancy or others within a stipulated period of time [4]. Malaysia continues to grow and had a rapid development of building stock. The government built on the initial stock of all categories of assets to fulfil the demand and providing quality services to the people through the provision of proper assets. The objective of the Malaysian government is to expand infrastructure facilities to keep abreast of the growing demand for infrastructure arising from the growth and transformation of the economy [23]. Infrastructure has received the largest allocation of public sector development expenditure in the Malaysian Plans.

In the First Malaysia Plan (1966–1970) the amount spent for the development was RM1, 39 million. Table 2.1 shows the allocation of development and expenditure spent in 2010–2015. The total investment by the Malaysia government was RM293 billion.

In the 11th Malaysia Plan, the RM260 billions of development expenditure represents a 16% increase from the previous plan, which bodes well for the growth trajectory of the construction industry. Buildings that are in poor state of performance will increase the rate of capital consumption. Durability of existing building components determines the period of their performance decay and, therefore, maintenance and refurbishment cycles which, owing to their occurrence, may have relevant effects on the environment and the economy [24]. The well-being of a country, therefore, depends on buildings because national building stock depends on the amount spent to maintain the buildings.

Many developed countries do not only spend more than half of their total investment in the construction industry to maintain their stock, but also invest about 10% of their GDP for maintenance of constructed facilities [23]. GDP from construction in Malaysia decreased to RM12, 572 million in the fourth quarter of 2016 from RM13, 062 million in the third quarter of 2016. GDP from construction in Malaysia averaged RM9, 680.29 million from 2010 until 2016, reaching an all-time high of RM13, 062 million in the third quarter of 2016 and a record low of RM6, 464 million in the first quarter of 2010.

Table 2.1 Yearly allocation spent for expenditure budget

No	Description	2010	2011	2012	2013	2014	2015	Total
1	Yearly expenditure budget	53,220	51,182	51,249	49,750	46,500	40,768	292,668

Source Financial Report of Ministry of Finance for the year 2010 until 2015

National Property Information Centre (NAPIC), kept a record of information pertaining to the building stock in Malaysia. The information on the existing inventories of the properties on a sectorial basis, namely residential, commercial, industry and leisure. According to NAPIC, the existing inventory (stock) are the units that have been issued with the certificate of fitness (CF) or temporary certificate of fitness (TCF). Table 2.2 shows the summary of existing stock of residential, shop, and industrial properties in H1 2016. While Table 2.3 shows the summary of existing stocks of the shopping complex and purpose-built office in H1 2016.

The Public Works Department (PWD) Malaysia has been given the mandate to record all government immovable asset since 2008. The main purpose of the asset registration is to be used as a basis of the information in the Malaysia Total Asset Management. The government's assets are listed in the Directory of Federal Government Premises Registration, (JKR21600-0012-12), 2011, published by the PWD Malaysia. The overall numbers of Government's building registered in Malaysia are 13,356 as shown in Table 2.4.

Table 2.2 The summary of existing stock of residential, shop, and industrial properties in H1 2016

State	Residential units	Shop units	Industrial units
WP Kuala Lumpur	421,201	25,441	5138
WP Putrajaya	6754	394	48
WP Labuan	11,761	856	560
Selangor	1,379,790	86,796	37,786
Johor	725,110	73,713	14,973
Pulau Pinang	402,494	31,032	8063
Perak	432,645	53,003	8004
Negeri Sembilan	243,530	19,135	5280
Melaka	167,676	18,828	6405
Kedah	284,657	26,749	3700
Pahang	240,185	19,470	3483
Terengganu	93,281	6668	864
Kelantan	64,824	9269	542
Perlis	24,094	3848	222
Sabah	170,531	17,531	5764
Sarawak	223,115	26,719	5621
Malaysia	4,891,648	419,452	106,453

Source Property stock report—NAPIC (2016)

Table 2.3 The summary of existing stock of shopping complex and purpose-built office in H1 2016

State	Shopping complex (s.m)	Purpose built office (s.m)
WP Kuala Lumpur	2,761,209	8,449,889
WP Putrajaya	61,090	2,052,944
WP Labuan	28,434	66,716
Selangor	3,216,322	3,307,385
Johor	1,709,220	1,083,867
Pulau Pinang	1,728,104	1,100,880
Perak	962,236	618,688
Negeri Sembilan	466,750	295,896
Melaka	433,023	398,487
Kedah	558,817	362,894
Pahang	272,203	397,067
Terengganu	128,325	342,721
Kelantan	287,447	381,462
Perlis	55,872	131,038
Sabah	678,884	685,975
Sarawak	803,087	674,288
Malaysia	14,151,023	20,350,207

Source Property stock report—NAPIC (2016)

Table 2.4 Categories of government buildings in Malaysia

No	Building categories	Total building registered
1	Office	690
2	Health	1250
3	Security	855
4	Infrastructure	138
5	Recreation	3
6	Religion house	20
7	Education	
	(a) Schools	10,070
	(b) University	161
	(c) Kindergarten	113
	(d) Ministry training centre	37
8	Industry	17
9	Cemetery	2
10	Commercial	0
11	Built up	0

Source Federal Government Premises Registration, PWD (2011)

2.4 School Buildings in Malaysia

2.4.1 Malaysian National Educational Systems

Olanrewaju and Aziz [25], reported that the education sector is very significant and integral part of the Malaysia Strategic Thrust. The Malaysian education structure can be divided into pre-tertiary and higher education levels. There are two governing authorities in the education sector. Pre-tertiary education (from pre-school to secondary education and teacher education) is under the jurisdiction of Ministry of Education Malaysia (MOE), while higher education is the responsibility of the Ministry of Higher Education Malaysia (MOHE). The level of education system in Malaysia can be divided into the following groups:

i. Preschool/Kindergarten education for children aged 4–6.
ii. Primary education from age 7 to 12 (Standard 1–6 for six years).
iii. Secondary education from age 13 to 17 (From 1 to 5 for five years).
    - Lower secondary from age 13 to 15 (From 1 to 3 for three years)
    - Upper secondary from age 16 to 17 (From 4 to 5 for two years) with the option to choose either academy secondary education; technical/vocational secondary. education; or religious secondary education
iv. Post-secondary education/Pre-university from age 18 (for one to two years) either From 6 (for one and half years) or matriculation certificate (for one year).
v. Higher education from age 18 onwards with certificate and diploma in polytechnics or colleges; education at Teacher Training Institutes; Bachelor's degree for three to five years; and Postgraduate studies (master's degree or Ph.D. studies, after acquiring a bachelor's degree for one to five years).

2.4.2 Categories of Educational Institutions

In the Malaysian education structure, there are seven broad categories of educational institutions which include both government and public-funded institutions and private institutions. There are five categories of school education in Malaysia:

i. Government and private sector kindergarten, which followed the National Preschool curriculum;
ii. Government schools that followed the Malaysian National Curriculum and National Examination;
    There are two categories of government-funded primary school, namely National schools where the medium of instruction is the national language and National-type schools where the medium of instruction is Chinese for Chinese National-type schools and Tamil language for Tamil National-type school.

Table 2.5 Student enrolment and number of public primary and secondary schools 2016

Categories of public primary schools	No. of schools	Student enrolment
Primary school	7772	2,685,403
Secondary school	2408	2,188,525
Total	10,180	4,873,920

Source EMIS data as 31 July 2016 and APDM data as 31 May 2016

iii. Government-aided schools that followed the Malaysian National Curriculum and National Examination;
iv. Private schools that followed the Malaysian National Curriculum and National Examination;
v. Foreign system schools which are International schools and expatriate schools that followed an international curriculum and use English as its medium of instruction while expatriate schools followed the national education system of the respective country using the respective national language of their country as the medium of instruction Table 2.5 shows the student enrolment and the number of public and secondary schools in Malaysia as of 2016. The total number of public schools are 10,180. Meanwhile the total number of schools registered until 2011 in asset registration were 10,070 as shown in Table 2.4. The different numbers in the EMIS data and government asset registration data is depended on the updated data in the government system. The latest updated data by government was published in 2011.

2.4.3 *Conditions and Performance of School Buildings*

School is to be used for a long period of time and is reported that it is in need of extra care [5]. School buildings designed in a variety of sizes, shapes and styles depending on the location, financial provisions, requirements specifications and design style of the building at the time [26]. Almost all school building in Malaysia designed in reinforced concrete structure (Federal Government Premises Registration, PWD, 2011). According to Gharpedia [27], reinforced concrete frame structure is less prone to corrosion apart from the simple construction. The school requires a proper management system to achieve maintenance objectives.

Structure of building will fail due to design error which affect the strength and stability of the building [28]. It is important to understand the property of material during design stage to minimise the defect during construction [29]. OSIM [28] identify defects can be involved in the deterioration of concrete structure building for example scaling, disintegration, erosion, delamination and many more. In order to achieve the effective maintenance for schools, an understanding of the maintenance management process is essential and crucial. School buildings are one of the important facility for getting everyday knowledge. The condition of schools is a very important factor in influencing the school environment and should be evaluated [30]. Mazlan et al. [31] stated that the poor building condition will affect the teaching and learning process.

The school building is the main asset to the learning process, so that the information on the current condition to the building is very important in managing to support school operations and providing a quality learning environment. Akasah and Alias [32], reported that effective school maintenance protects capital investment, ensure the health and safety of the children and support educational performance. There are many cases which involved component failures and maintenance of Malaysian school building in the late of year 2007 until 2010 as shown in Table 2.6.

The latest incident was the roof of SMK Seri Nilam which collapsed due to decay structure on 13th May 2014. Based on the research by Yaakob et al. [33], 80% of Malaysian schools faced several problems at age 15 years and above, 71% school education has no systematic work system in the maintenance management and only 25% stated said maintenance action taken less than 1 month to solve the maintenance. Meanwhile Akasah and Alias [32] states that one of the problems is the process in getting maintenance funding and 70% of Malaysian school community does the maintenance activity by themselves without using maintenance funding.

2.5 Types of Building Defects

Ahzahar et al. [9] describes the term of defects briefly as imperfection; fault; a blemish or deficiency. According to Drabold et al. [35], defect is nonconformity of a component with a standard or specified characteristic. Research by Khalid and Othuman Mydin [36], reported that defects can be referred to fault on something that detract from perfection, whilst building damage can be seen when any structure, material, equipment and also element of the building was not fully functional. A defect is a shortfall in performance occurring at any time in the life of the product, the element or building in which it occurs (BRE digest 286) or defect is interpreted as a fault in an element, material or component of a building.

Defects can occur at construction or during the handling over or occupancy stages. A study by Chong and Low [37], have found that defects found during construction and occupancy were usually different even the defects had similar descriptions. The defects had different roots and were of the different nature. Chong and Low [37] also mentioned that building defects often appear after 2 years of the first age of the building or after occupancy. While building defects that occur after 6 or 7 years considered the effects of the adoption. Besides defects can also occur and looks over a period longer that cannot be estimated.

Patent defects are defect that covers all types of defects that can be seen by the naked eye. For example, walls cracked, clogged drains, leaky roofs, doors and windows that could not be closed tightly and so on. Meanwhile latent defects are a not known defect or is only detected when the building was occupied. According to Chong and Low [29], design deficiencies are the major contributors to latent defect

Table 2.6 Report on maintenance and school building condition issues

No	Title	Issues	Date	Source
1	The school building is critical	6% from 9642 schools in Malaysia are categorized in critical physical condition	30.08.2010	Bernama
2	Roof collapse	Half the roof of SMK Wakaf Tapai collapsed due to support failure	13.08.2010	Utusan Malaysia
3	Ministry of education goes for school physical condition auditing	MOE will audit the physical condition of school buildings for upgrading the physically bad condition of school building especially in rural areas	02.04.2010	Berita Harian
4	5 years old school building cracked	5 years old school building of Sekolah Kebangsaan Taman Bersatu cracked due to sedimentation, instability of building site and problematic of roof structure	30.3.2010	Berita Harian
5	Danger of sick building to students	A Tamil school (SJKT) in Glenmarie is confronted with the sick building issue that exposes students and teachers to the risk of accidents	15.12.2009	Utusan Malaysia
6	Education ministry needs RM500 million to repair 567 schools, teachers quarters and hostels	RM500 million for the repair work for 567 schools that had been categorised as 'very critical condition' throughout the country	11.09.2009	Bernama
7	The school roof collapsed	The roof structure of the school porch was slightly imbalanced caused by a hit from one of the machines	19.07.2009	Utusan Malaysia
8	The investigate cause school wall collapsed	School wall collapsed at School Datuk Harun Selangor	27.09.2008	Bernama
9	Roof frame collapsed	The roof frame of new three-storey Sekolah Menengah Kebangsaan (SMK) Sect. 16 which is still under construction collapsed due to weakness of the roof structure	29.08.2008	Berita Harian
10	School hall closed	The roof structure of the school hall was believed to fall due to the truss joint to support the roof is not strong	14.04.2008	Berita Harian

(continued)

Table 2.6 (continued)

No	Title	Issues	Date	Source
11	Obliged to contribute RM145 to PIBG SK Taman Tun Dr. Ismail	The parents were shocked when they must cover the PIBG fee, including PIBG fees of RM40, RM15 hall fees, RM20 for maintenance fees and RM70 for computational cost	15.01.2008	Utusan Malaysia

Source Hamzah et al. [34]

and can only prevented by improving the design. The defect of this type usually cannot be detected or known to the naked eye, especially by those who are not so skilled about the construction of the building. Besides that, this defect can pose a hazard at any time or something that cannot be known or predicted accurately at an early stage. Other terms used for the building defects as follows:

(a) A Defect is a problem caused by deficiency in either design, specification or workmanship
(b) Damage is a problem caused by us in the use of building, impact, abrasion and chemical spillages
(c) Deterioration is a problem caused by the worsening of the material.

2.6 Common Building Defects and Possible Causes

There are several building defects which usually occur in building parts such as roofs, walls, floors, ceiling, toilets, door and window. Ahzahar et al. [9] stated that common types of building defects include cracks, faulty electrical wiring, faulty plumbing, faulty drainage systems, faulty ventilation, cooling or heating systems, inadequate insulation, inadequate fire protection, mold, dry rot, wood rot, termite, etc. According to the PWD Malaysia guideline [38], the defects were listed and summarize according to the building element and possibly caused to the defect. Table 2.7 shows the defects and its possible causes for elements of the building.

2.6.1 Defects in Buildings

Defects are something that has flaws that can reduce the value of the building. Defects occur in different extends and forms in all types of buildings. The occurrence of defects in buildings are due to the factors which contribute to building defects. The list of defects is explained in the following sub chapter.

Table 2.7 Summary of building defects and its possible causes

Defect/element	Possible causes
(a) Leakage	
Roof	(1) Old roof had been damaged (2) Nail hole had been loose (3) Inaccurate roof installation not according to the specification (4) The roof structure is damaged, broken or rotted (5) Roof position leaning or slanting (6) Failure of waterproofing (7) Failure of material (8) Shoddy workmanship (9) Lack of maintenance
Gutter	(1) Clogged with dust, rubbish, dirt, leaves and others (2) Out of date/old (3) Failure of connection (4) Inappropriate gutter size (5) The number and failure of bracket (6) Lack of maintenance
Drain	(1) Broken drain by stress loads, (2) Soft base land-slip, (3) Failure of cement mortar (4) Lack of maintenance
Water tank	(1) Tank ball is damaged and not functioning (2) Rusty tank (3) The burden of excessive water (4) Failure of tank connector (5) Slanting tank due to damage of tank structure/broken (6) Lack of maintenance
Cold water and waste pipe	(1) Out of date/old (2) Load pressure from the top (3) Connectors failure, broken (4) Shoddy workmanship (5) Inaccurate installation (6) Clogged pipes (7) Lack of maintenance
(b) Tilt/crook	
Door frame	(1) Unbalance load transferred (2) Inaccurate installation (3) Settlement (4) No lintel (5) Lack of maintenance
Window frame	(1) Unbalance load transferred (2) Inaccurate installation (3) Settlement and no lintel (4) Lack of maintenance

(continued)

Table 2.7 (continued)

Defect/element	Possible causes
Timber frame and connector	(1) Distorted due to weather (2) Thin frame (3) Lack of maintenance
Building fixture	(1) Misused (2) No insulation or paint (1) Weather exposure, such as heat and rain (2) Lack of maintenance
Board/plank	(1) Distorted due to weather (2) Screws or nails is not enough (3) Lack of maintenance
Metal frame	(1) Stress or high attraction (2) Lack of maintenance
(c) Rust/corrosion	
Roof connection and finishes	(1) Weather exposure, such as heat and rain (2) Out of date/old (3) Lack of maintenance
Metal frame including handrail, balcony, balustrade, door and window grille	(1) Non-coated with insulation or painted (2) Chipping paint (3) Exposed to high humidity or water acid or salt (4) Contact with faeces of animals or birds (5) Misused (6) Lack of maintenance
Building fixture	(1) Misused (2) No insulation or paint (1) Weather exposure, such as heat and rain (2) Lack of maintenance
Gutter and rain water down pipe	(1) Out of date/old (2) Unpainted and peel off paint (3) Lack of maintenance
(d) Peel off	
Wallpaper	(1) Out of date/old (2) Vandalism (3) Dampness (4) Lack of maintenance
Paint on wall, ceiling, door and window frame	(1) Old paint (2) Inappropriate new layer of paint over the old paint (3) Climate change (4) Inaccurate painting work (5) Lack of maintenance

(continued)

Table 2.7 (continued)

Defect/element	Possible causes
(e) Mold/fungi	
Toilet floor	(1) Leakage of waste pipe (2) Wet floor (3) Clogged pipe (4) Shoddy workmanship, porosity (5) Lack of maintenance
Floor	(1) Leakage near pipeline under floor (2) Lack of maintenance (3) Wet floor (4) Water ponding (5) Lack of maintenance
Door, window and fascia board	(1) Weather exposure, such as heat and rain (2) Poor/low quality of building material (3) Fungal and insect attack (4) Lack of maintenance
Kitchen floor	(1) Sink leakage (2) Waste pipe and Water supply pipe leakage (3) Lack of maintenance
Skirting board	(1) Wall dampness (2) Ground floor dampness (3) Insect attacks (4) Lack of maintenance
Carpet	(1) Water seepage through the floor (2) Pipe leakage, from near water supply, waste pipe, soil pipe (3) Rain sprinkles (4) Lack of maintenance
Pàrquet floor	(1) Water seepage through the floor (2) Rain sprinkles (3) Pipe leakage, from near water supply, waste pipe, soil pipe (4) Poor/low quality of building material (5) Insect attacks
Brickwall	(1) Poor/low quality of building material (2) Weather exposure, such as heat and rain
Floorboard	(1) Roof leakage (2) Water seeps through soil to the wall plates (3) Insect attacks
Timber threshold	(1) Weather exposure, such as heat and rain (2) Gutter leakage or overflow
Timber beam, crossbeam	(1) Pipe leakage (2) Rain sprinkle (3) Poor/low quality of building material (4) Insect attacks

(continued)

Table 2.7 (continued)

Defect/element	Possible causes
(f) Dampness	
External and internal wall	(1) Water seeps through soil (2) Poor workmanship and Poor installation (3) Failure of waterproofing (4) Pipe leakage near wall (5) Overflow water from drain (6) Insufficient plaster (7) Flooding and Water ponding
Centre of external wall	(1) Rain sprinkles (2) Pipe leakage from above floor (3) Leakage of rain water down pipe (4) Insufficient plaster
Top of external wall	(1) Rain sprinkles (2) Gutter leakage (3) Overflow water (4) Insufficient plaster
Ceiling	(1) Roof leakage (2) Pipe or tank leakage (3) High condensation from air conditioning
Floor	(1) Water seeps through soil (2) Failure of waterproofing (3) Water ponding (4) Insufficient of floor gradient (5) Clogged waste pipe (6) Pipe leakage
Floor and wall near toilet	(1) Pipe leakage—cold water, waste pipe (2) Clogged pipe (3) Water ponding
Wall above ceiling	(1) No flashing (2) Insufficient plaster (3) Firewall outside wall
Floor and wall basement area	(1) Waterproofing (2) Insufficient plaster (3) Pipe leakage and high-water level
(g) Sagging	
Floor	(1) Poor/low quality of the material (2) Too wide of beam distance (3) Overload
Tank based	(1) Small size—inappropriate size (2) The distance of the crossbar too wide (3) Dampness due to the leakage (4) Insufficient number of the crossbar
Roof trusses	(1) Insufficient number of trusses (2) Inappropriate size of the trusses (3) Overload (4) Dampness due to roof leakage (5) Low quality of material

(continued)

Table 2.7 (continued)

Defect/element	Possible causes
(h) Settlement	
Ground floor	(1) Soft foundation (2) Not enough compaction (3) Not enough reinforcement (4) Near vibration (5) Overload
Foundation	(1) Soil erosion (2) Drain leakage or broken (3) Tree root taken out/thrown (4) Soil movement and near vibration
Structure and column	(1) Insufficient number and size of foundation (2) Thin foundation (3) Imbalance of soil compaction (4) Insufficient number of piling (5) Near vibration
(i) Detached/uprooted/dislodged	
Porch	(1) Movement of foundation (2) No connection/joint/reinforcement (3) Overload
Roof	(1) Insufficient collar or purlin (2) Damage of roof structure (3) Roof to distort (4) Poor or shoddy workmanship (5) Poor installation (6) Roof movement due to external pressure (7) Strong wind
Flashing	(1) Inappropriate size (2) Wrong installation (3) Strong wind (4) Not enough nail
Handrail	(1) Pressure (2) Wear and tear (3) Loose screw (4) To sway
Plastering or rendering	(1) Too dry during rendering/plastering (2) Too thick and near vibration (3) Out of date/old (4) Vegetation root
(j) Condensation	
Space above ceiling	(1) Insufficient heat protection of roof (2) Less ventilation (3) Difference in temperature (4) Exposed water tank above ceiling
Wall and two-layer window	(1) Failure of sealant (2) Manufacture fault

(continued)

Table 2.7 (continued)

Defect/element	Possible causes
Store	(1) Near dampness wall (2) Insufficient ventilation
Space or room in basement	(1) Insufficient ventilation (2) Dampness wall (3) Wet floor (4) Water ponding
(k) Crack	
Ground floor	(1) Settlement (2) Less reinforcement (3) Load pressure (4) No wide connection
Upper floor	(1) Load pressure (2) No wide connection (3) Failure of plaster (4) Vibration (5) Structure movement (6) Broken beam (7) Wide floor
Flat roof	(1) Less reinforcement (2) Less or no wide joint/connection (3) Change of weather (4) Load pressure
Wall beside door and window	(1) No lintel (2) Inappropriate size (3) Failure of door and window frame
Wall joint major crack	(1) Difference in temperature between new and old wall (2) No exmet or connector between two walls (3) Failure of joint
Wall plaster minor crack	(1) Inaccurate mixed of plaster (2) Brick wall too dry (3) Shoddy workmanship
Load bearing wall major crack	(1) Settlement (2) Soil movement (3) Weak foundation (4) Near vibration (5) Overload (6) Vegetation root (7) Mortar failure
Structure wall major crack	(1) Crack at the lower beam (2) Bent or broken (3) Horizontal load pressure (4) Building structure slanting (5) Vegetation root (6) Less stiffeners (7) Wall too wide

(continued)

Table 2.7 (continued)

Defect/element	Possible causes
Structure, beam crack	(1) A design fault (2) Overload (3) Settlement (4) Vibration (5) Structure movement
Apron or balcony crack	(1) Settlement (2) Erosion (3) No reinforcement connection to the structure of building (4) Vegetation root (5) Broken drain (6)Load pressure
Other vertical structure crack	(1) Loading (2) Less reinforcement (3) Movement of below structure
(l) Wrong installation	
Wrong wiring	(1) Wrong installation (2) Extra installation over the limit (3) Not accordance to the specification
Water hammer	(1) Vertical pipe from tank to higher tank (2) The used of bigger size of pipe (3) No balancing
Toilet door frame broken	(1) Inappropriate height of stone heel
Water ponding or overflow	(1) Inappropriate height of floor drop (2) Inappropriate gradient of the floor (3) Clogged water trap
Plant vibration	(1) Loose screw (2) Inappropriate plant based
(m) Insect attack/biological agent	
Timber stairs, roof, door and window structure, timber panel, timber ceiling and others	(1) Termite (2) Insect (3) Mouse/mice
Palm roof	(1) Mouse/mice (2) Bird (3) Snail
Floor and ceiling	(1) Bad (2) Swallow bird nest
External and balcony ceiling	(1)Swallow bird nest
(n) Clogged	
Distribution pipe	(1) Dregs (2) Dented pipe (3) Lack of maintenance
Soil pipe	(1) Rubbish or dirt (2) Dented pipe or broken in soil (3) Lack of maintenance

(continued)

Table 2.7 (continued)

Defect/element	Possible causes
Rain water down pipe	(1) Leaves and rubbish (2) Dented pipe (3) Lack of maintenance
(p) Dirty/unclean	
Finishes, wall, floor, other	(1) Lack of maintenance (2) Poor design

Source JKR 21602-0004-13 (2013)

2.6.1.1 Leakage

Leakage is the escape of water through a hole or crack [39]. Recent studies identified water leaks as the most common defects in buildings and is the major cause of early onset of corrosion and concrete deterioration [40]. Water and moisture ingress are often the source of defects and damage [41]. The failure of sanitary fittings, inadequate maintenance of the sewer piping system and defects to waterproofing membrane can cause dampness and seepage inside the buildings [10].

Water leakages in buildings result mainly due to inadequate construction design, inferior materials, poor workmanship and deterioration of building materials [40], poor ratio of concrete mix, no proper vibration while concreting, concrete not being dense, poor water curing, poor selection of concrete raw materials, poor workmanship, damages caused to the building by various other agencies, construction, expansion and extension not comply with specification [39]. Chong and Low [29], reported that major defects for plumbing and sanitary, roof and ceiling defects are pipe leakage and water seepage [42]. Figure 2.1 shows pictures of different extent of defects due to leakage.

Leakage: Rainwater pipe — Leakage: Roof and ceiling — Leakage: Floor and ceiling

Fig. 2.1 Leakage to the building elements. *Source* PWD inspection reports

Corroded iron railing Corroded window frame Corroded door knob set

Fig. 2.2 Corrosion to the building elements. *Source* PWD inspection reports

2.6.1.2 Tilt/Crook

Crook is innate defects caused by the natural characteristics of wood to shrink or expand in response to water vapour in the air and it can lower its strength, durability and usefulness, and caused decay [43]. According to Ramly [1], tilt or slant is due to unbalance load transferred, inaccurate installation, and settlement. This defect happened usually in timber frame including door and window frame, board or plank and also to metal which is due to stress or high attraction. Figure 2.2 shows pictures of form of crook/tilt defects.

2.6.1.3 Rust/Corrosion

Rust or corrosion is one cause of poor durability of reinforced concrete and this is due to chloride or a carbon dioxide attack that reduce the passivity of the reinforcement in the alkaline concrete system [44]. The process of basic corrosion is part chemical and part electrical and it requires a combination of moisture and oxygen and the typical causes of corrosion as follows:

i. Uncoated steel, steel that has not been coated to suit with the environment;
ii. Not being properly maintained;
iii. Lack of a vapour barrier and/or adequate insulation inside the building;
iv. Unaddressed maintenance issues such as leaky roofs, plumbing leaks, standing water, etc., that lead to chronic exposure to moisture; and
v. Incorrect design/construction of the building foundation.

Figure 2.2 shows pictures of defects due to the corrosion.

2.6.1.4 Spalling

Spalling occurs subsequent to cracking and it will become huge and the concrete on which account, becomes spalled. Spalling can occur due to the following:

i. Rusted steel reinforcement.

Fig. 2.3 Spalling to the building elements. *Source* PWD inspection reports

ii. When the steel reinforcement in the component structure is rusting, it will crack along that reinforcement. When this cracking continues, the concrete spalling follows suit.
iii. Extreme weather exchange.
iv. Weather is the main factor to concrete spalling. This is because when the weather is hot, concrete expands and when the weather is cold, concrete shrinks. The extreme weather exchange causes the concrete to expand and shrink rapidly and eventually causing the concrete to spall if left unchecked.
v. The concrete building that is exposed to the chemical agent/material is more easily to decay. This is because most chemical matters have high erosion factor. The slow erosion will contribute to the concrete spalling, if not remedied.
vi. Chemical reaction.
vii. The inferior quality of concrete.
viii. The quality of the concrete would be compromised if the ratio of the concrete mixture does not comply with the standards. The strengths of the concrete will become weak if the component of the structure is damaged.
ix. Unsuitable reinforcement or overloaded.
x. Concrete is weak at the tension load. The use of the steel reinforcement can support the load, but the use of unsuitable steel reinforcement can make concrete crack and spall. It is because if the reinforcement cannot support the tension load resulting in the concrete structure cracking and spalling.

Figure 2.3 shows pictures different extent of defects due to the spalling.

2.6.1.5 Peel off/Torn/Tattered/Ragged

Peeling paint usually occurs on building facades, on plastered walls, columns and other areas which are exposed to excessive rain and dampness [45]. It appears in a wide range of geometries [42]. Peeling off either for paint or wallpaper is due to exposure to sunlight, rain, wind and dampness [42] which can easily turn the surfaces of paint to be chalky and wrinkled or blistered [45].

Peeling off paint at wall　Peeling off paint at column　Peeling off paint at stiffener

Fig. 2.4 Peel off to the building elements. *Source* PWD inspection reports

Peeling paint results due to poor surface preparation and moisture penetration through walls due to improper application of waterproofing membranes around pipe penetrations, joint and walls [46]. Chong and Low [29], reported that peeling paint is one of major external wall defects which are caused by weather, moisture from wet areas, and door defects due to vandalism, and impacts by occupants or loads. Building located near the sea faced a much greater risk once the signs of peeling paints are visible on the exterior walls. Figure 2.4 shows the pictures the peel off paint defects in the building.

2.6.1.6 Mould/Fungi

During the past decades, mold and fungi in building have become a growing concern. The growth of mold and fungi caused by excessive dampness [42], moisture problems, and poor ventilation [47]. Molds and fungi grow when gained moisture and nutrient [48].

This happens when dust and dirt penetrate small holes, opening and cracks in the walls and mortar joint. The growth of mold fungi indoors deteriorates the air quality, influences human health [42] and causes economic losses. Figure 2.5 shows pictures of mold defects in the buildings.

Mold at wall　Mold / fungi at stair　Mold at wall

Fig. 2.5 Mold/fungi to the building elements. *Source* PWD inspection reports

Table 2.8 Cause of decay

Type	Agent	Environmental factors
Biological	Fungi (dry rot, wet rot, molds and others) bacteria; actinomycetes; lichens, mosses and algae wood-boring insect larvae (wood-worm, death watch beetle and others) carpet beetle, moths, book lice and silverfish termites	Moisture and humidity, air movement, temperature, light, dust and food source
Chemical	Acids, alkalis and solvents	Pollution remedial treatment
Physical	Mechanical abrasion, general handling and others, decomposition by physical agents such as prolonged heating, fire and moisture	Normal use, visitor wear, accidental damage, sunlight, heating, fire, damp
Radiation	Ultraviolet light	Exposure to sunlight

Source Wong et al. [47]

Door rot Wall rot Floor rot

Fig. 2.6 Rotting to the building elements. *Source* PWD inspection reports

2.6.1.7 Decay

Decay of timber in building is caused by a broad spectrum of bacteria, fungi and insects [47] and the extent of the damage depends on materials and the condition [47]. Table 2.8 shows the cause of decay in buildings. Figure 2.6 shows pictures of different extent of defects due to the decay.

2.6.1.8 Dampness

Dampness are particularly subjected to water. According to Othman et al. [49], cited from (World Health Organization (WHO), 2009), moisture problem is "any visible, measurable or perceived outcome caused by various leaks of water". Moisture it is a natural movement and major sources of defects in building components and elements [50]. According to Douglas [51], moisture are water present in stuffs while humidity is the amount of water vapor present in the atmosphere. They also stated that moisture

Dampness at wall | Dampness at wall partition | Dampness at ceiling

Fig. 2.7 Dampness to the building elements. *Source* PWD inspection reports

of more than 70% relative humidity can damage the structure, fabric, component and element of the building including risks to human health. Almås et al. [52] stated that 76% major cause of the defect is due to moisture in the building. Meanwhile WHO, identified that moisture caused 76–80% of building envelope defects. Water entering a building through penetration such as capillary action through the crack between joints and trap the moisture behind renders.

The main causes of the dampness in building are due to rain, condensation, rising damp, flooding, services leaks, construction process, use of the building and moisture in the air. While Othman et al. [49] identified the common moisture problem in buildings from her study as follows:

(i) Rainwater or ground water leaking into the enclosure;
(ii) Plumbing leaks and spill;
(iii) Water wicking from capillary suction;
(iv) Rainwater, condensation or plumbing water;
(v) Infiltration of warm or moist outside air;
(vi) Exfiltration of warm or moist indoor air;
(vii) Insufficient dehumidification by heating, ventilation and air conditioning system;
(viii) Unvented or poorly vented sources;
(ix) Poor condensate drainage;
(x) Enclosure of wet materials in building during construction.

Figure 2.7 shows pictures of different extent of defects due to the dampness at different building element.

2.6.1.9 Sagging

The roof sagging is a common problem found in both old and new buildings and it can be seen when viewing the roof from the exterior and it appear wavy and not straight. The implication of sagging lead to roof collapse [53]. Awol [42], reported that sagging is one of the defects found in his study, especially for building with pitch roof which bulge downward. Sagging occurred due to load to the existing structure

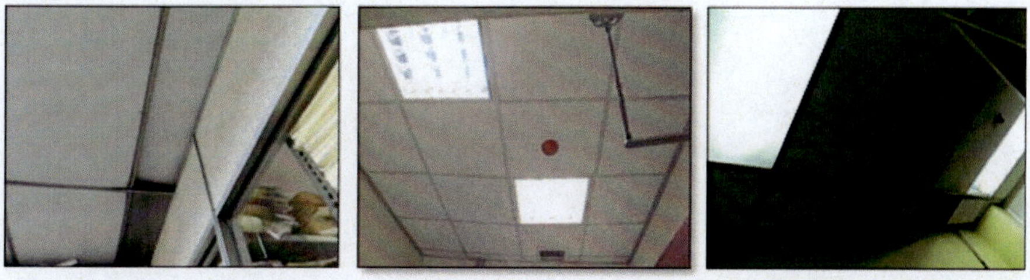

Sagging ceiling Sagging ceiling Sagging ceiling

Fig. 2.8 Sagging to the building elements. *Source* PWD inspection reports

such as wind, sheathing and roof materials [53]. Structural problems can be due to improper design or construction or due to environmental factors such as age, damage to the roof's structure element or failures of the other supporting system including wall and foundation [54]. Figure 2.8 shows pictures of ceiling sagging in the building.

2.6.1.10 Crack

Most buildings will develop cracks at some point in time of their service life [55] due to the poor mixing of building materials. The occurrence of crack due to the material age, particularly exposed to moisture as they expand and contract alternately. According to Ugwu et al. [55], cracks may appear in the wall brick/block, mortar, plaster, beam-wall joints, columns-wall joints, slabs, beams, columns, foundations and in many other places. Research by Chiaro et al. [56] stated that the cracks are classified into structural and non-structural categories while Kashyzadeh and Keshch [57] identified the structural ones are due to faulty design, faulty construction or overloading.

Meanwhile, non-structural cracks are due to internally induce stresses. The cause of occurrence of cracks is due to moisture change, thermal variations, elastic deformations, creep and foundation movement and settlement of soil [56], failure design, change service, use of poor quality and make mistakes in construction technology [57].

There are varieties types of vertical crack such as vertical crack away from corners, at the centre of the panel, and between bay window corners, horizontal crack and diagonal crack [56]. Figure 2.9 shows pictures of crack at different areas of building elements.

2.6.1.11 Clogged

Clogged are a common problem in both residential and commercial buildings. Clogged in the building refers to sink and also drains. Clogging caused by construction spoils, inappropriate slope at execution and bad habit of the user [58]. A clogged

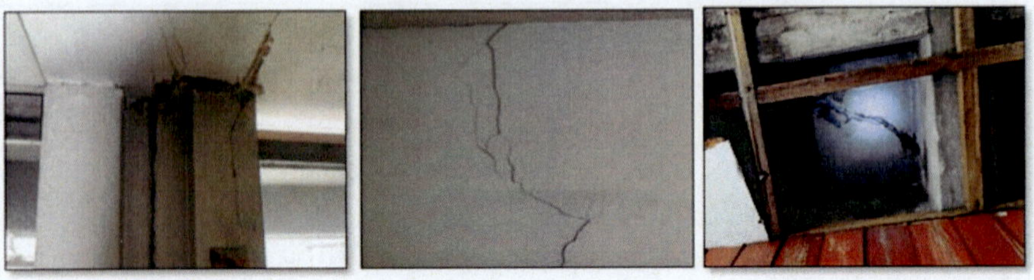

Crack at column Crack at wall Crack at floor

Fig. 2.9 Crack to the building elements. *Source* PWD inspection reports

Clogged rainwater downpipe Clogged perimeter drain Clogged floor trap

Fig. 2.10 Clogged to the building elements. *Source* PWD inspection reports

drain result when dirt, skin flakes and especially hair bind to soap scum on the walls of drain pipes and will block the passage of water and reduces the water flow. This situation will lead to an overflow from the affected plumbing fixture. Meanwhile clogged sink is due to kitchen grease and oil cake onto drain pipe walls. Anything that comes from food or is used for cooking potentially caused grease clogged pipe or due to mixture of calcium and fat [59]. Common causes of blocked drain are silt and building debris, leaves, tree roots, faulty sewer lines human error by placing foreign objects in the drain. Figure 2.10 shows pictures of clogged defects in the building.

2.7 Impact of Building Defects

Built problems within the buildings are solely a government responsibility, especially for the government buildings and remain liable for any accident that might occur. Defects is a building flaw that reduces the value of the building and can cause a dangerous condition of the building and also affect the building appearance. According to Pereira et al. [60], the impact of building defects are to social and economic impact which include the depreciation of assets, increased maintenance cost, litigation costs, disharmony and increased stress levels.

Various research shows that the impact of the building defects can cause to the increment of the maintenance cost [61, 62]. Inefficiency in handling the defects will generate several effects and negative impact to the user and the owner of the buildings. Milion et al. [63] has proven that the defects caused a negative impact on user satisfaction. Meanwhile Costanzo [24] noted that the defects gave impacts to health, safety and security, functionality and efficiency of technical and space elements and psychological, social, cultural and aesthetic performances.

References

1. Ramly, A. (2006). *Panduan Kerja-kerja Pemeriksaan Kecacatan Bangunan*. Hizi Print Sdn Bhd.
2. Vitelli, V., & Nelson, D. R. (2004). Defect generation and deconfinement on corrugated topographies. *Physical Review E*, *70*(5), 0151105. Publisher's Version.
3. Zakiyudin, M. Z., Fathi, M. S., Rambat, S., Tobi, S. U., & Rejab, N. (2014). An overview of building maintenance management in Malaysia. In *7th IRERS IBIMA International Conference*. Selangor.
4. Rusli, E. A., Mydin, M. A., & Abdullah, S. (2014). Study on significant mechanisms of maintenance culture among public building users in Malaysia. In *SHS Web of Conferences* (Vol. 11, pp. 1–6). EDP Sciences.
5. Ahluwalia, S. S. (2008). *A framework for efficient condition assessment of the building infrastructure*. University of Waterloo.
6. Kyle, B. R., Vanier, D. J., & Lounis, Z. (2002). The BELCAM project: A summary of three years of research in service life prediction and information technology. In *International Conference on the Durability of Building Materials and Components* (pp. 1–10). NRCC. Retrieved from http://www.nrc.ca/irc/ircpubs
7. Shittu, A. A, Adamu, A. D., Mohammed, A., Suleiman, B., Isa, R. B., Ibrahim, K., & Shehu, M. A. (2013). Appraisal of building defects due to poor workmanship in public building projects in Minna, Nigeria. *IOSR Journal of Engineering (IOSRJEN)*, 30–38.
8. Hermawan, F. (2013). Toward sustainable practice in building project at indonesian local government: A case study of construction failure and building failure (defects) in Central Java Indonesia. In *Proceeding the 6th Civil Engineering Conference in Asia Region: Embracing the Future Through Sustainability*, (pp. 26–33). Jakarta, Indonesia.
9. Ahzahar, N., Karim, N., Hassan, S., & Eman, J. (2011). A study of contribution factors to building failures and defects in construction industry. In *Procedia Engineering. The 2nd International Building Control Conference* (Vol. 20, pp. 249–255). Elsevier. https://doi.org/10.1016/j.proeng.2011.11.162
10. Wong, J. T. Y., & Hui, E. C. M. (2005).Water seepage in multi-storey buildings. *Facilities*, *23*(13/14), 595–607.
11. Mohd Isa, H., Hassan, P., Mat, M. C., Isnin, Z., & Sapeciay, Z. (2011). Tracking architectural defects in Malaysian hospital projects. In *IEEE Symposium on Bussiness, Engineering and Industrial Applications (ISBEIA) Langkawi* (pp. 298–302).
12. Hamilton, B. (2003). Maintenance of building important. Retrieved from http://www.hba.org.my/news/2001/601/maintenance.htm
13. Hamilton, B., & Wan Salleh, M. (2001). Maintenance of building important (pp. 22–27).
14. Ali, A. S., Keong, K. C., Zakaria, N., Zolkafli, U., & Akashah, F. (2013). The effect of design on maintenance for school buildings in Penang, Malaysia. *Journal Structural Survey*, 194–201.
15. Roslan, T., Ahmad, A. G., Zakaria, N., & Sulieman, M. Z. (2014). Assessment of factors affecting building maintenance and defects in public building in Penang Malaysia. *Architecture Research*, 48–53. https://doi.org/10.5923/j.arch.20140402.03

16. Olanrewaju, A. A. (2012). Quantitative analysis of defects in university buildings: User perspective. *Built Environment Project and Asset Management*, 167–181.
17. Minato, A. T. (2003). Representing casual mechanism of defective design: a system approach considering errors. *Construction Management Economic*, 297–305.
18. Adnan, H., Fauzi, Z. F. M., Rahmat, I., Supardi, A.. (2012). Maintenance management for public infrastructure for Malaysian local authorities. 7: 514–1522.
19. Zawawi, E. M., Kamaruzzaman, S. N., Ali, A. S., & Sulaiman, R. (2010). Assessment of building maintenance management in malaysia: Resolving using a solution diagram. *Journal of Retail and Leisure Property*, 349–356.
20. Pintelon, L. (1999). Information technology: Opportunities for maintenance management. *Journal of Quality in Maintenance Engineering, 5*(1), 9–24.
21. Che Ani, A. I, Sairi, A, Johar, S., Razak, M. Z., & Tawil, N. (2015). The practice of building inspection in Malaysia total asset management. *Jurnal Teknologi*, 1–11.
22. Abu Backer, M. M. A., & Wan Yusoff, W. Z. (2014). A review of asset management in Malaysia local government. In *Proceedings of the 25th International Business Information Management Association Conference—Innovation Vision 2020: From Regional Development Sustainability to Global Economic Growth, IBIMA 2015* (pp. 3577–3582).
23. Naidu, G. (2008). Infrastructure development in Malaysia. In N. Kumar (Ed.), *International infrastructure development in East Asia—Towards balanced reional development and integration.* ERIA Researh Paper Project Report 2007-2, Chiba IDE-JETRO, pp. 204–227.
24. Costanzo, E. (2006). Building defects: Survey and impact over sustainability. *WIT Transaction on the Built Environment*, 361–368. https://doi.org/10.2495/ARC60361
25. Olanrewaju, A., & Aziz, A. A. (2015). *Building maintenance processes and practices.* Springer Science and Business. https://doi.org/10.1007/978-981-287-263-0-4
26. Mustafa, F. A., & Rafeeq. D. A. (2019). Assessment of elementary school buildings in Erbil city using space syntax analysis and school teachers' feedback. *Alexandria Engineering Journal, 58*(3), 1039–1052. ISSN 1110–0168. https://doi.org/10.1016/j.aej.2019.09.007
27. Gharpedia. (2018). https://gharpedia.com/blog/why-steel-is-used-in-reinforced-concrete/
28. OSIM. (2008). *Ontario structure inspection manual. St. Catharines.* Ministry of Transportation.
29. Chong, W. K., & Low, S. P. (2006). Assessment of defects at construction and occupancy stages.*Journal of Performance of Constructed Facilities*, 213–221.
30. Mahli, M., Che-Ani, A. I., Abd-Razak, M. Z., Tawil, N. M., & Yahaya, H. (2012). School age and building defects: Analysis using condition survey protocol (CSP) 1 matrix. *International Journal of Civil, Environmental, Structural, Construction and Architectural Engineering, 6*(7), 56–58.
31. Mazlan, R. R., Mohammed, A., & A, B. (2008). Public school maintenance: Competency issue. In *2nd Engineering Conference on Sustainable Engineering Infrastructure Development & Management*. ENCON2008-244.
32. Akasah, A. Z. A., & Alias, B. M. (2009). Development of the generic maintenance management process modelling for the preservation of heritage school building. *International Journal of Intergrated Engineering*, 43–53.
33. Yaakob, M., Faiz, M., & Yunus, J. (2016). The relationship between school culture and professional learning community in Malaysia. *International Journal of Academic Research in Business and Social Sciences., 6*, 231–236. https://doi.org/10.6007/IJARBSS/v6-i12/2489
34. Hamzah, N., Mahli, M., Che Ani, A. I., Tahir, M., Abdullah, N., & Tawil, N. (2010). The development of smart school condition assessment based on condition survey protocol (CSP) 1 matrix: A literature review. *Science, Engineering and Technology*, 620–625.
35. Drabold, D. A., & Estreicher S. K. Defect theory: An armchair history. In D. A. Drabold & S. K. Estreicher (Eds.), *Theory of defects in semiconductors. Topics in applied physics* (Vol. 104). Springer. https://doi.org/10.1007/11690320_2
36. Khalid, M., & Othuman Mydin, M. A. (2012). Building condition assessment and defect analysis on heritage shophouses in Penang, Malaysia: Case studies. *International Journal of Engineering*, 441–444.

37. Chong, W. K., & Low, S. P. (2005). Latent building defects: Causes and design strategies to prevent them. *Journal of Performance of Constructed Facilities*, 283–289.
38. PWD. (2013). *Building condition inspection for existing building.* Malaysia Public Work Department.
39. Panchal, D. J., Shah, N. H, Sindhav, C. R, Joshi, C., Chauhan, A. (2015). Waterproofing challenges and suggested remedial measures for high rise buildings: A case study. *IJSRD—International Journal for Scientific Research & Development, 3*(10).
40. Choudhry, R. M., Gabriel, H. F., Khan, M. K., & Azhar, S. (2017). Cause of discrepancies between design & construction in the Pakistan construction industry. *Journal of Construction in Developing Countries, 22*(2), 1–18.
41. Lam, E. S. (2007), *Water seepage, flood and hydraulic.* Paper presented at Structure Congress, 16–19th May, 2007, Unites States.
42. Awol, A. (2016). Assessment on causes of defect and the maintenance management practices on low cost building (a case study of jimma town condominium). *International Journal of Engineering and Technical Research,* 151–156.
43. Georgiou, J., Smith, J., & Peter E. D. (2014). A benchmark study on housing defects in Victoria, Australia. *International Journal of Construction Management*, 37–41.
44. Zaki, A., Chai, H. K., Aggelis, D., & Alver, N. (2015). Non-destructive evaluation for corrosion monitoring in concrete: A review and capability of acoustic emission technique. *Sensors., 15*, 19069–19101. https://doi.org/10.3390/s150819069
45. Bakri, N. N. O., & Mydin, M. O. A. (2014). General building defects: Causes, symptoms and remedial work. *European Journal of Technology and Design,3*(1), 4–17.
46. Chew, M. Y. L., Kang, H. L., & Wong, C. W. (1998). *Building facades: A guide to common defects in tropical climates.* World Scientific, 20 Feb 1998, Architecture, 80 pages.
47. Wong, A., Singh, A., & Ling, W. W. C. (1996). Major decay types recognized in an extremely durable timber species belian (Eusideroxylon zwageri) from East Malaysia.
48. Pheng, L. S., & Wee, D. (2011). Impact of ISO 9000 on the reduction of building defects. *Architectural Science Review*, 367–377. Retrieved from http://tandfonline.com/loi/tasr20
49. Othman, N. L, Jaafar, M., Wan-Harun, W. M., & Ibrahim, F. (2015). A case study on moisture problems and building defects. *Procedia-Social and Behavioral Sciences*, 27–36. Seoul, S.Korea.
50. Bagdiya, N. V., Wadalkar, S. (2015). Review paper on construction defects. *IOSR Journal of Mechanical and Civil Engineering (IOSR-JMCE)*, 88–91.
51. Douglas, J. (1996). Building performance and its relevance to facilities management. *Facilities, 14*(3/4), 23–32.
52. Almås, A.-J., Lisø, K. R., Hygen, H. O., Øyen, C. F., & Thue, J. V. (2011). An approach to impact assessment of building in a changing climate. *Building Research & Information, 39*(3), 227–238.
53. Wardhana, K. (2003). Study of recent building failures in the United States. *Journal of Performance of Constructed Facilities,* 151–158.
54. Schmidt III, R., Mohyuddin, S., & Austin, S. (2008). Using DSM to redefine buildings for adaptability. Paper presented at 10th International Design Structure Matrix Conference, Stockholm, Sweden.
55. Ugwu, O. O., Arop, J. B., Nwoji, C. U., & Osadebe, N. N. (2013). Nanotechnology as a preventive engineering solution to highway infrastructure failures. *Journal of Construction Engineering and Management, 139*(8), 987–993.
56. Chiaro, G., Kiyota, T., Pokhrel, R. M., Goda, K., Katagiri, T., & Sharma, K. (2015). Reconnaissance report on geotechnical and structural damage caused by the 2015 Gorkha Earthquake, Nepal. *Soils and Foundations, 55*(5), 1030–1043.
57. Kashyzadeh, K. R., & Kesheh, N. (2012). Study type of cracks in construction and its controlling. *International Journal of Emerging Technology and Advanced Engineering, 2.*
58. Rodriguez-Gonzalez, L., Miriyala, A., Rice, M., Delgado, D., Marshall, J., Henderson, M., & Ergas, S. J. (2020). A pilot-scale hybrid adsorption-biological treatment system for nitrogen removal in onsite wastewater treatment. *Journal of Sustainable Water in the Built Environment, 6*(1), 04019014.

59. Balkan Sewer. (2015). Balkan sewer and drain cleaning. https://www.balkandraincleaning.com/wp-content/uploads/call-a-pro.jpg
60. Pereira, C., de Brito, J., & Silvestre, J. D. (2020). Harmonised classification of the causes of defects in a global inspection system: Proposed methodology and analysis of fieldwork data. *Sustainability, 12*(14), 5564.
61. Siddiqi, K., & Hunt, A. (2015). *Construction claims: Impacts of defective construction on long-term facility maintenance costs*. Faculty Publication, paper 3658. Retrieved from http://digitalcommons.kennesaw.edu/facpubs/3658
62. Allotey, S. E. (2014). *An evaluation of the impact of defects in public residential buildings in ghana. Civil and Environmental Research, 6*(11), 58–64. Retrieved from http://www.iiste.org
63. Milion, R., Alves, T., & Paliari, J. (2016). Impacts of defects on customer satisfaction in residential buildings.

Chapter 3
Factors Contributing to Building Defects

Abstract This chapter reviews the literature related to the factors that contributing to the building defects. Most of the factors contribute to the building defects and it is impossible to control all the factors above during design, construction and occupancy stage. These factors are also related to each other to cause building defects. In this study, it is found that the factors identified have some impacts towards the building condition. These factors are determined as independent variables in this study. The relationship between factors and building condition has been studied and analysed in this study. To identify the types of defects and the root cause of the problems in the actual buildings in this study, it required building condition assessment (BCA) and find the cause of problems and provide accurate information of the building defects. BCA as a method of inspection for data collection was analysed and reported in the next chapter.

3.1 Introduction

This chapter is to describe the factors which contribute to the building defects. This chapter provided the literature studies and analyses which covered the aim of gathering information and knowledge based on the objectives studied. This chapter presented the phenomenon defects which consists of the various factors which contribute to building deterioration toward condition of the building. The factors that lead to building defects are arranged according to the group of similarity.

3.2 Defect Factors

Building defects can lead to many building problems [1]. The building problem is due to numerous main factors [2]. According to Ramly [3], defects in the building are the problem faced by most of the building regardless of building construction techniques or age, but it depends on the causes and the factors causing the occurrence of

S. Yacob et al., *Managing Building Deterioration*, Management in the Built Environment,
https://doi.org/10.1007/978-981-16-5860-0_3

the defects. To specify the defect/damage of the actual building, it requires inspection services from the experts to find the cause of problems and provide accurate information for repair work. There are various factors that contribute to the building defect or deterioration. Previous studies have reported that the factors which cause the building defects are tabulated in Table 3.1.

From the literature, it shows that the defects factors are continuously discussed to overcome the building defects issues. According to Chong and Low [22], defects are the key trigger in the construction industries. Different types of facilities constructed will generate different types of defects. Meanwhile the study by Ahzahar et al. [23], examined that most of the studies concerns only on overall construction defects and only a few studies about defects in relation to each building element, i.e., architectural, structural, mechanical and electrical, etc. Building defects are categorized by their consequences. Those that caused defects are different depends to the building design, type, location, and use of the building. Previous studies have confirmed that all factors were studied within various types of building such as school, multi storey building, office building, hospital, bridges, apartment, steel structure building, hostel, mosque and others.

Table 3.2 shows that the summary of factors to cause defects from the previous studies. The table shows that a total of 40 factors found caused the building defect. The factors of design and maintenance is the highest factors found by researchers that contribute to building defect/deterioration. The design factor seems to occur during pre-occupancy or during construction or for a new building which fall under technical issues. Meanwhile maintenance falls during occupancy stage or for existing building. The material also will lead to the building defects. The research by Chong and Low [22], reported that defects due to the material depends on time to develop and will appear after acceptance of buildings or during occupancy compare to during construction.

In this study, all factors will be analysed and to confirm certain factors which contributed to defect especially for school buildings. The factors have been arranged according to group based on similarities. In order to adopt the 29 factors used in this study, a semi structured interview with an expert was conducted to confirm and agree that those factors are the common factors that contribute to the defects especially for building in Malaysia, according to the literature studied. Table 3.3 shows group of factors in the generation of defects.

3.3 Technical Factor

Josephson and Larsson [24], stated that one of factors that contributed to the defects in the building is technical factors. These technical factors, whether lead to the pattern defects or latent defects. Choudhry et al. [25], noted that a technical error occurs when a person fails to correctly carry out procedures. According to Josephson and Larsson [24], the unique technical design is one of the reason building fails and lead

Table 3.1 Factors contributing to the building defects

No.	References	Deterioration factors
1	[4, 5]	Studies have confirmed that poor workmanship is the main factors contributing to maintenance defects 1. Poor workmanship
2	[6]	Eight Contributing Factors to Building Defects 1. Climatic conditions 2. Location of the building 3. Construction materials 4. Building type and change in use 5. Maintenance of building 6. Faulty design 7. Corruption 8. Lack of supervision
3	[7]	Factors affecting building condition 1. Facility condition index 2. Type of building and size 3. Demographic factors of occupant, age, gender, background 4. Age or major year of renovation of building 5. External and internal conditions 6. Advance in information technology 7. Maintenance frequency
4	[8]	The author investigated and found that a total of 13 factors contribute to defect 1. Age of building (year) 2. Waterproofing system selection and details 4. Plumbing details 5. Access for plumbing maintenance 6. Selection of fitting 7. Fitting details and layout 8. Maintenance of fittings 9. Ventilation 10. Material durability (year) 11. Material performance 12. Usage 13. Maintenance practices
5	[9]	A total of 6 factors were found as important factors for building deterioration 1. Building ownership status 2. Element quality 3. Age 4. Type of paint 5. Responsible and element for building maintenance 6. Type of material and location
6	[10]	Six factors of building deterioration that contribute to the building failures 1. Design 2. Detailing 3. Construction 4. Maintenance 5. Material 6. External

(continued)

Table 3.1 (continued)

No.	References	Deterioration factors
7	[11]	The author reported that a total of 17 factors found to affect the building deterioration a. Agent Related to the inherent quality characteristic 1. Workmanship 2. Materials 3. Storage 4. Transport 5. Manufacture 6. Site management b. Environment 7. Ventilation 8. Condensation 9. Elevation of building 10. Micro-environment condition 11. Traffic emissions 12. Weathering factors c. Operating Conditions 13. Accessibility for maintenance 14. Quality and frequency of maintenance 15. Wear and tear 16. Category of users 17. Mechanical impacts
8	[12]	Factors affecting building defects 1. Action of load underestimated at design and construction stage 2. Insufficient knowledge of construction/fixing of building element/component 3. Failure to carry out necessary maintenance or incorrect identification of the cause of the defect 4. Failure joint element
9	[13]	Thorough examination of the cause of defects found as follows: 1. Biological 2. Meteorological 3. Geochemical 4. Natural Hazard 5. Human intervention 6. Pollution 7. Mismanagement 8. Inappropriate use 9. Poor maintenance and remedial work
10	[14]	Studies have found that poorly specifications and unclear design lead to building defects 1. Poorly worded specifications and 2. Unclear designs often lead to lower construction quality

(continued)

Table 3.1 (continued)

No.	References	Deterioration factors
11	[15]	Eleven major groups of defects through literature and interviews showed that defects were generated from the following: 1. Civil design, 2. Architectural design, 3. Design issues on maintenance practicality and adequacy, 4. Consultant firm administration and staff, 5. Construction drawings, 6. Contractor administration, 7. Construction materials, 8. Construction equipment, 9. Specification
12	[16]	Research by the author has shown that a total of 12 factors causes of deterioration in buildings 1. Human 2. Chemical 3. Atmospheric 4. Structural 5. Moisture 6. Fire 7. Faulty design 8. Faulty construction 9. Faulty materials 10. Faulty systems 11. Cleaning 12. Vandalism
13	[17, 18]	The authors found that a total of 10 factors contributing to building defect 1. Weather 2. Environmental conditions 3. Soil impact 4. Poor design 5. Chemical attacks 6. Structural movement due to poor structural design 7. Installation method 8. Workmanship 9. Maintenance issues 10. Site working conditions
14	[18]	Six factors that affect building deterioration 1. Mechanical agent 2. Electromagnetic agents 3. Thermal agents 4. Chemical agents 5. Biological agents 6. The building user
15	[19]	Studies by the authors shown that the factors caused defects as followed: 1. Chemical attack 2. Poor workmanship

(continued)

Table 3.1 (continued)

No.	References	Deterioration factors
16	[20]	The author conclusively has shown that 4 factors cause to building defects or deterioration 1. Faulty design 2. Operation and installation 3. Poor material and system 4. Unexpected user's requirements
17	[21]	The author pointed that there are 5 factors that cause to defects 1. Wear and tear 2. Poor Design decision 3. Material 4. Workmanship

to the defects. Chong and Low [26], reported that design, specification, detailing and layout were significantly contributing to building defects as technical design factor.

3.3.1 Faulty Design

Faulty in design or poor design can generate defect that is latent in nature. Research by Chong and Low [26], reported that faulty design will become the platform for latent defect. This is due to the conditions where the building element will be exposed to the defects during occupancy. Georgiou et al. [27] revealed that design error is a problematic issue in construction. The above findings are consistent with the findings of past studies by Josephson and Larsson [24] and Assaf et al. [15] which mentioned that all defects caused by the client, and the design and defects were mostly generated by design. The authors also pointed that faulty design also due to poor detailing such as:

Insufficient allowance for expansion or contraction:

(i) Absence of weathering
(ii) Poor jointing between different materials and components
(iii) Poor Specification
(iv) Lack of adequate consideration for future maintenance
(v) Inadequate provision for access to carry out maintenance activities.

Examples of defects caused by faulty design are poor detailing, improper design for plumbing, drainage, and others. Ahzahar et al. [6] in his research stated that faulty design by reducing the size of the columns, the size of reinforcement bars and foundations are the most common design error in construction. This situation possibly leads to the uncertainty situation in the future, sometimes result of misjudgement, and leading to assumption decisions. Previous study by Josephson and Larsson [24] emphasized lack of information about the requirement of the project may cause a lack of understanding of how a specific task fits within the project and could lead to

Table 3.2 Summary mapping of the factors

No.	Factors	Total
1	Age	3
2	Atmospheric	1
3	Biological agent	1
4	Building orientation	1
5	Building location	2
6	Building ownership	1
7	Building Size	1
8	Chemical	3
9	Climatic condition	4
10	Condensation	1
11	Construction	3
12	Design	12
13	Detailing	3
14	Drawing	1
15	Electromagnetic-agent	1
16	Faulty systems	1
17	Fire	1
18	Human	3
19	Information technology	1
20	Lack of knowledge	1
21	Lack of supervision	1
22	Maintenance	12
23	Material	11
24	Mechanical agent	2
25	Moisture	1
26	Operation and installation	2
27	Soil Impact	1
28	Specifications	2
29	Structural	1
30	Thermal agent	1
31	Type of building	2
32	Usage	1
33	User	4
34	Vandalism	1
35	Ventilation	2
36	Waterproofing	2
37	Wear and tear	2

(continued)

Table 3.2 (continued)

No.	Factors	Total
38	Weather	2
39	Working condition	2
40	Workmanship	3

Table 3.3 Grouped of factors in generation of defects

Technical	Environmental	Human
(1) Faulty design (2) Poor detailing (3) Faulty drawing (4) Not complying with the specification (5) Structural	(1) Insect attack (2) Reaction of biological agent (3) Change of climatic condition (4) Poor Ventilation	(1) Misuse by user (2) Vandalism (3) Wear and tear (4) Change of usage (5) Poor workmanship (6) Lack of supervision
(6) Poor waterproofing (7) Improper use of material 8) Lack of maintenance (9) Poor construction (10) Building size (11) Type of building (12) Building orientation	(5) Reaction of thermal agent (6) Excessive moisture (7) Reaction of chemical agent (8) Soil movement impact	(7) Lack of knowledge (8) Poor operation (9) Lack of cleaning

a design error. This statement also supported by Salim et al. [28] in her study that major building design failure developed building defects.

3.3.2 *Poor Detailing*

Conflicting detailing exists in the design drawing. Conflicting occurs between civil drawings or mechanical and electrical drawings. According [5] design drawings are prepared separately for each section, e.g. civil drawings or mechanical drawings, without consulting other related parties. Besides that, some of the drawings lack of detail for example, structural elements, joints, electrical connection and plumbing work. This action may cause conflicts and left to resolve situation especially to the contractors. The example of the defects caused by poor detailing are spilling of brickwork, horizontal cracks, random cracks near wall, and etc.

3.3.3 *Faulty Drawing*

Defects are one of the major causes of disputes and construction litigation due to drawing discrepancy during construction. Lack of detail references in the drawing lead the contractor to execute the construction based on his understanding and judgement. Others issue for drawing factors are discrepancies in drawing and bill of quantities in the contract, drawing uncompleted, unclear specification and poor drawing by unskilled draftsmen. Talib [29], pointed that what appears good on drawings or on the computer screen is sometimes difficult to build which lead to the defects in construction.

3.3.4 *Not Complying with Specification*

The study by Assaf et al. [15], found that contractors, non-compliance with specifications tend to do things on their own way and few follow the specifications. As a result, all the effort spent during the design stage is worthless. Unclear or poor specification indicated in the drawing also lead to generate defects and lead to the construction defects.

3.3.5 *Structural*

Structural factors mean any defect for the structural element and this is due to poor design, faulty workmanship, poor maintenance, overloading and materials. Defects occurred due to these factors such as cracks (intrinsic, corrosion-induced, load-induced and settlement), spalling, exposed reinforcement, delamination, water leak and tilt. Defects to the structure of the building due to reaction of the structural element to settlement, moisture, shrinkage, thermal, loading, natural aging, reaction to corrosive element in the weather condition and deterioration due to inadequate inspection and maintenance.

3.3.6 *Poor Waterproofing*

Waterproofing problem is a common issue in the building which leads to the moisture problems [30]. Waterproofing factors will cause to defect such as water leaks, tilt, algae/fungus attack, mold, and mildew growth to the element of the building. The problem involving waterproofing always occurred at basement, flat roof, floor at toilet and walls. According to the research by Suffian [31], the problem that most associates with the flat roof is a waterproofing issue. Due to this problem, the condition of the building will deteriorate and will increase the cost of maintenance. Common damage to the waterproofing at flat roof is when the membrane layer will harden and caused it dislodged from its original position [32]. This is due to direct rays of the sun which cause the layer to shrink and expand during rain. The damage is also caused by poor workmanship of installation, weather, wear and tear, daily uses and materials used.

3.3.7 Improper Use of Material

Material defects usually appear after the construction period or during the occupancy stage. According [30] quoted from the study by Chew [8], selection of materials in terms of durability is most important in order to control the occurrence of building defects. Ahzahar et al. [6], reported that faulty construction tends to lead contractor to construct the project using the sub-standard of materials, not according to the specification and without the permission and concern of client and consultant awareness. Besides that, according to Assaf et al. [15], the faulty material can lead to defects and decay to the finishes, elements, components and services of the building.

The faulty material arises due to:

(a) Improper selection or use of materials;
(b) Thermal movement in different materials;
(c) Improper selection of materials unsuitable for climatic conditions;
(d) Use of non-durable materials;
(e) The use of expired materials;
(f) Poor handling and storage of material;
(g) Use of sub-standard material;
(h) Inconsistent and inadequate mixing of material on the site;
(i) Inadequate storage facilities on site; and
(j) Inadequate inspection of materials by supplier or receiver or contractor.

3.3.8 Lack of Maintenance

Adequate and systematic building maintenance can prevent building defects to occur [33]. Talib [29], reported that in his study, the first important factor affecting building defects are lack of maintenance. Some maintenance defect arises and will be expected as buildings age and more apparent during occupancy of the building. Without a proper and comprehensive maintenance will lead to the worst building condition or structural failures [33] and increase the cost of maintenance [34]. In addition, failure to carry out necessary maintenance will lead a building defects in the buildings. Furthermore, [15] reported that failure to proper planning, carry out the routine maintenance, and failure to promote awareness of maintenance needs to the occupant or to who use the building can causes building defects and deterioration.

According to Ahzahar et al. [6], pointed that the building which lack of maintenance can generated several defects and lead to the structural failures, future financial burden, pose legal and etc. It is important to secure the stability and lifespan of a building. So that it is important to do the regularly inspection to the main structural elements, walls and roofs and other common building problems or areas [33]. It is important that buildings continue to be properly maintained to ensure that they can effectively function [6].

3.3.9 Poor Construction

Insufficient knowledge of construction/fixing of building element/component will occur a building defect to the building. According to Assaf et al. [15], defects due to construction are as below:

(i) Inadequate measurement by contractor which is under or over-measures the size of a building element, the location and the ratio of the materials;
(ii) Use of damaged formwork which can affect the quality of the construction work. For example, by producing a honeycomb or porous concrete surface;
(iii) Improper method of construction involved structure and all elements of the building. For example, excavation or piling work next to the existing building, usually causes settlement or vibration to the foundation of the buildings;
(iv) Applying paint in suitable condition and surfaces will cause continuous maintenance action and it turns to major action if it is performed improperly;
(v) Applying inadequate waterproofing and drainage by unskilled workers or inexperienced contractor can result water seepage through the element of building such as ceiling, floor, wall, beam and columns;
(vi) Insufficient concrete cover to reinforcement will lead to defects such as spalling, exposed rebar, and cracks;
(vii) Failure of designer/architect/or engineer to monitor work in progress;
(viii) Improper construction of cold joint, especially for concrete and plastering work. This involved when to construct or to joint between new concrete and old concrete. If not treated properly, will cause surface cracking and water seepage;
(ix) Loss in adhesion between materials when the material is used in the wrong place or wrong temperature which results the material life has expired or;
(x) Early formwork removal can cause permanent deflection and cracking of the structure; and
(xi) The inadequate curing process can cause corrosion or loss of adhesion between cement plastering and the surface of the concrete.

Table 3.4 Building categories and description

No.	Building categories	Description
1	Operation	Used to prepare the Department's core services (e.g. Police Station, Hospital, Military, School)
2	Side	Used as the support functions (e.g. Library, Training Camps, Multi Car Park, Administration block)
3	Does not operate	Unused and abandoned (e.g. Building waiting for disposal)
4	Control	Administered on behalf of the Government (e.g. Buildings that have historical and cultural heritage

Source [35]

3.3.9.1 Building Size

The size of the building also can affect the occurrence of defects in the building. Without a proper maintenance and supervision, the size and the number of the defects will increase due to the complexity of the buildings. The size of building project has become complex, which lead to many defective designs resulting delay and poor quality in construction.

3.3.9.2 Type of Buildings

The type and the function of the building will affect the defects in the building. Different type of building and organization, operation will lead to different size and the number of the defects in the building. Building that changes their use and space will effect of the new functionality of the existing structure [6, 33]. As highlighted by Ahzahar et al. [6], the change in use of building means to accept for the certain load and the installation of new and extra services such as air conditioning, fittings and etc. This situation may lead the defect to occur.

According to the PWD Malaysia guideline [35], the building categories were listed according to its function and type of building operations. Table 3.4 shows the building categories and the function of the building which will affect the defects in the buildings.

3.3.9.3 Building Orientation

Building orientation provided a building with passive thermal and visual comfort. It is referred to the way a building is situated on site and the positioning of window, roof and other features [36]. Orientation of the building will affect the condition of the building. Ling et al. [37], mentioned that the building orientation related to the wind flow over and around the building will create negative and positive pressure regions. The defects may occur such as roof tiles can move under wind action, tiles may blow away or drop into guttering. Optimum orientation of building significantly contributes to eliminate the access of sunrays or reaction of thermal which can caused serious defects to the buildings [38].

3.4 Environmental Factor

The environmental factors, including insect attack, biological agent, climatic condition, ventilation, thermal agent, moisture, chemical agent and soil impact. According to Hassanain et al. [16], the categories of environmental factors such as wind, rain, sun, and pollution, chemical which interacted with corrosion environment, climate, excessive of moisture in the atmosphere.

3.4.1 Insect Attack

Research by Ridout [39], reported that money is spent on repairing the damage due to timber decay, insect and small bird attack every year. This is due to beetle attack, ants, bees and termites. Insect also is one of causes of defects such as decay of timber and disastrous losses in timber building. Termites are extremely destructive to timber and create severe damages to the wooden structure. They eat out and inside the wood, leaving only the thin shell of wood. According to Johar et al. [40], in some cases, termites can cause damage to concrete foundations.

The research study by Tan and Corlett [41] also mentioned that termites can cause damage to wood and concrete particularly by species from tropical regions. Termite can be divided into two types which is dry wood and subterranean. According to Pellitteri [42], subterranean termites use the wood as food which have a source of moisture, usually from soil, a leaking basement water pipe. They normally feed inside the wood, leaving a thin outer layer as a protective barrier against water loss. Meanwhile ants and bees like to breed and build nests instant of wood. Ants make small holes on the wood surface. Bird can create defects by building their nest at the building.

3.4.2 Biological Agent Attack

According to Johar et al. [40] biological agent is well known for their disastrous attacks. Fungi is a group of organisms that can lead to decay and discoloration of wood [43]. Fungi can be divided into several types which is brown rot and white rot. According to Che Ani et al. [44], timber that is at below 20% moisture content is immune to fungal attack; meanwhile the timber above this level, possibly easy attack by fungi. This statement also supports by Johar et al. [40], fungi can be detected with attending of moisture. Rot attack a damp wood and penetrate from the surface and causes defect to the wood elements and turning the original colours of the wood. March et al. [43], stated that fungi are a group of organisms and can cause decay and discoloration to the building material especially wood.

Biological agents can be divided into four categories [1] as follows:

(i) Surface growth which include bacteria, fungi, algae, lichen and mosses;
(ii) Insect vermin which is the predominant damage or deterioration of materials;
(iii) Animal vermin is the pests which harm building materials such as rats, mice and small bird. According to Salim et al. [28], birds create nesting, creating holes in the timber buildings also bring moisture and aesthetic defects with their droppings.
(iv) Plant agents—vegetation which cause the major damage to the building, such as tree roots disrupting foundations and penetrating underground drains. Others are plants growing in gutters and blocking them.

3.4.3 Changes of Climatic Condition

The climatic condition will affect the building material and the condition of the building [6]. It is crucial to consider the climatic condition and the effect on building materials [33]. It involved reaction of the structure, external fabric, finishes, and etc. Climatic condition such as wind, rain, sun, dust or pollution in the air. Every structure is vulnerable to the wind, rain, flood and hot condition [33].

Extreme and rapid temperature changes exacerbate exterior problems by causing expansion and contraction of the building. The example of defects due to climatic condition such as leaks in and around metal window frames and will cause damage to the wall and to interior spaces. The element can rust and affect the integrity of exteriors such as spalling at floor slab, column, beam, exposed rebar, and fungi attack. According to Assaf et al. [15], example of defects due to climatic condition are cracked due to heat, salt reaction, and etc.

3.4.4 Poor Ventilation

Ventilation in the building is necessary and the function of the ventilation is to control humidity. This factor interrelated with moisture factor and both will cause a building defect to happen. According to Chew [8], poor ventilation can affect the structure of the building due to air and moisture to flow and enter through the building envelope. Besides that, [45] mentioned that poor ventilation also leads to mold growth. Poor ventilation will affect the humidity level in the building to increase either it is due to natural ventilation, mechanical ventilation or mixed mode ventilation. High indoor humidity also can cause mold growth and bacteria. This also support with the research by Institute of Medicine (US) Committee on Damp Indoor Spaces and Health [46], high indoor humidity can cause mold growth and bacteria.

Lack of adequate ventilation can prohibit proper airflow, leading to condensation issues and results in mold and mildew which lead to water damage, fungus, wood rot, etc. [33]. Mold growth caused by the condensation when the dew point of the air above the temperature of the surface and water vapour will condensate that cold surface. Condensation happens due to the different temperature between indoor and outdoor temperature.

3.4.5 Reaction of Thermal Agent

Othuman Mydin et al. [1] stated that temperature is particularly relevant to components that are exposed to an unobstructed sky, for example roofing, cladding and external structural members. The actual temperatures reached can lead to either temporary or even permanent changes in physical or chemical properties, such as embrittlement at low temperature and accelerated oxidation at high temperature. Changes of the temperature are also relevant when assessing the consequences of thermal expansion and contraction such as stresses within materials when changes of size are restrained and strained imposed on jointing materials when components are free to change the size [1].

3.4.6 Excessive Moisture

Moisture can cause building defects and commonly happens at every building [14]. According to Ahmad [33], moisture caused 75–80% of building envelope defects. Meanwhile research by Almas et al. [47], found that 76% major building defect in the building are caused by moisture. Moisture can be from humidity, leak or faulty plumbing, irrigation, excessive moisture in the air/atmosphere which may lead to condensation and corrosion.

Besides that, moisture which penetrate into the external wall, giving dampness to the wall and create a suitable condition for fungi growth. Defect caused by moisture such as dampness due to roof leaks, efflorescence leaks through cracks, moisture between walls caused paint peeling, bulking, plumbing leaks, leaking ceiling and etc. According to Chong and Low [22], the presence of moisture triggered several moisture related defects such as poorly installed waterproofing and leaking pipes. According to Chew [8] the main building moisture problems due to roof, wall and ceiling leaking and he also found that 14 major defects in walls and floors are water leakages through the cracks, pipe penetration and joints.

3.4.7 Reaction of Chemical Agent

According to Johar et al. [40], defects by chemical agents cause physically coarse conditions and brittleness. Causes of defect or deterioration in the building by a chemical agent promoted to corrosion. Ridout [39] stated that exposure to alkaline environments has caused the timber loses strength and softened, and its physically coarse conditions and brittleness. Interaction of chemical agent with certain material or component causing disintegration. The example of defects caused by chemical factors are discoloration due to concrete age, alkali aggregate reaction, and etc. [40], highlighted that moisture will fasten the decay process and allowing decay agent to reproduce in wood especially when exposed to high level of moisture.

3.4.8 Soil Movement Impact

One of the major problems in the movement of soil is the presence of water flow [48]. Soil impact factor will lead to settlement due to vibration from excavation or piling work. Besides that, poor soil compaction also leads to settlement of the building. According to Assaf et al. [15], improper soil compaction work tends to soil settle for a certain time, and it will settle at a later stage and cause settlement. According to Suffian [31], soil settlement is critical to the building and infrastructure especially in swampy areas. His study also found that the most affected areas are building aprons, retaining wall, perimeter roads and fencings.

Othuman Mydin et al. [1] in his study, reported that soils such as clay are possibly to make a big movement due to its natural moisture absorbent and liquid can freely flow. The moisture changes in earth particle leads to changes in load and affect the stability of the soil underneath the building. The effects of the movement will cause a movement which lead to building cracks to wall and floor [1].

3.5 Human Factor

Human error or factors has proven to be problematic issues within the building industry and most failures are caused by human [49]. Human error also can occur in all human activities across an organization. This included mistakes in construction projects maintenance, user, supplier and others. Haan [49] cited that human failure can be divided into three categories:

(i) Error of knowledge (ignorance);
(ii) Error of performance (carelessness and negligence); and
(iii) Error of intent (greed).

Through his study also human error is due to failure to perform the task, incorrect performance of the task, extraneous acts, sequential errors, and failure to perform within the allocated time. Research by Samah et al. [50], reported that causes of many defects in the building are associated with human factors. The above finding is consistent with the findings of past studies by Jorgensen [51] who stated that the defects in the building are caused by human factors such as lack of knowledge, information, motivation and negligence of construction workers. Besides that, poor or shoddy workmanship, user, vandalism, wear and tear, usage, operation and cleaning are also caused by human factors. Meanwhile, [16] reported that human factors are due to failure to clean, poor maintenance, ignorance of the causes of defects, and adopting a negative attitude.

3.5.1 Misuse by User

Defects do occur as a result of action by the building occupants [1]. Condensation caused by the used of equipment such as air conditioning is a well-known example. Many buildings built misused by the occupant or user. A defect caused by intended misuse due to lack of knowledge on the correct method of usage of building element or systems. Horner et al. [52] identified improper use of building and delay reporting of failure by user was significant contributor to building defect and maintenance. Meanwhile [53] confirmed that normal use of building by user results in natural wear and tear and this leads to further deterioration of building condition.

3.5.2 Vandalism

According to Assaf et al. [15], one of the causes of defects or deterioration in building is vandalism. This is due to lack of security, attitude of the occupant, failure to promote awareness among occupants of the consequences of vandalism, incorrect selection of material and finishes in circulation area which are prone to vandalism and poor maintenance or failure to maintain and repair area of damage by vandals thus encouraging more vandalism. Vandalism is the factors that caused the defects on the building components. Chong and Low [22], prove that softer building materials often invite vandalism.

3.5.3 Wear and Tear

The definition from Wikipedia, wear and tear is a damage that naturally and inevitably occurs as a result of normal wear or aging. Olanrewaju [54], noted that the definition of wear and tear is a combination of two separate words. According to Coboulid Advanced Learner's English Dictionary (2006), wear and tear are the damage or change that is caused to something when it is being used normally. According to

Bakri and Mydin [2], the building material can be deteriorated easily because of wear and tear. A defect in a building occurs through general wear and tear [55] and also can occur over time due to wear and tear [56]. In other words, wear and tear are depreciated, reduction or fall in the functional performance or value of a building or services, which arises as a result of normal or fair use [54].

3.5.4 Change of Usage

Building that change their use and spaces should consider the effect of the new use on the existing structure because the buildings were built to only hold certain load [6]. The building which have been converted and change the function of the building will affect the condition of the building. In order to fulfil the new requirement of the building, the renovation of the building involved the installation of new air conditioning system, wiring, and services affects the appearance of the building. Without proper installation, and supervision will lead to building defects.

3.5.5 Poor Workmanship

Normally workmanship problems are due to faulty of contractors [30]. A previous study has shown that poor workmanship as one of the factors that can lead to the building defects [6]. The workmanship is the skill and quality put into making a product. Poor workmanship can lead to building defects or failures to the structure. Poor workmanship is also due to negligence and inadequate quality control during the execution of construction work either during or post construction especially during rectifying, replacement or maintenance work. If workers are careless or do not follow proper protocol, it can end up with a finished product that lacks the quality anticipated. Chong and Low [22], quoted a study from Josephson and Larsson [24] found that the most defects due to human factors are caused by forgetfulness and carelessness. Suffian [31], found that workmanship issues associated with small contractors which are not well trained or lack of skills.

Poor workmanship may result to the various building defects such as corrosion, mold, plumbing issues, damaged electrical wiring, injuries and death, loss of water through evaporation, cracks in the foundation and walls and leaking roofs. Examples of poor workmanship are installing products and materials incorrectly, incorrect water to cement ratio, improper plumbing, and poor concrete compaction. According to Chong and Low [22] in their research found that defects caused by poor installation methods including poor mixing of materials, poor handling of materials, poor planning that lead to poor completed quality, failure to provide proper joints, gaps or materials to avoid defect.

3.5.6 Lack of Supervision

Ahzahar et al. [6], stated that the major influence on the project performance is due to the quality of site supervision. This support by statement of [57], stated that lack of supervision will lead to the major causes of rework constant maintenance. Adequate supervision certainly reduced most defects and leading to low maintenance during operation [15, 58]. Agwu [59], stated that building failure is due to poor or lack of supervision. Lack of supervision during construction or replacement/renovation work. Glewwe and Muralidharan [60], mentioned that, well-trained supervisors have important role in minimising the amount of rework due to defects.

3.5.7 Lack of Knowledge

In general, defects are considered to arise as a result of lack of knowledge, lack of information or lack of motivation [22]. Most lack of knowledge issues is due to the limited knowledge of the contractors or workers. For examples is failure to understand the drawing and the method of work and lack of communication between parties involve. Research by Suffian [31] shows that the workers with limited knowledge will result in the repair approach for the area is not so good and will result in problems to recur. Waziri [57], stated that it is important to deliver the building as specification to the desired quality through competent workforce. The workforce must be experienced and technically sound to realize the project objectives [58].

3.5.8 Poor Operation

Operation of the building is the factors that contribute to the building defects [30]. The different operating hour in the building contributes to the moisture in the building especially hospital. Building operation is one of the factors that contribute to the defect problems [30]. Lack of operation leads the building to unhealthy building. Proper building operation is important to support the maintenance of the building [61].

3.5.9 Lack of Cleaning

Cleaning or housekeeping is to keep the building maintained, cleanliness, tidiness, upkeep and smooth running [62]. Without a proper cleaning of all building components result to an expensive defect to the building [28]. Lack of cleaning can cause defects such as water ponding due to clog at floor trap, leakage at gutter and insect and pests transmit diseases. In other words, [28] also reported that without proper cleaning, hazards can be created especially to the safety issues such as slipping on greasy, wet and dirty surfaces, tripping over loose objects on floors, staircase, platform and many more.

3.6 Selection of Independent Variables

In this study, the independent variables will be used according to the literature. Under technical defects were design, detailing, drawing, specification, structural, waterproofing, material, maintenance, construction, building size and type of building. Environmental factors consist of insect, biological agent, climatic condition, ventilation, thermal agent, moisture, chemical agent, and soil impact. Meanwhile, causes of human factor such as user, vandalism, wear and tear, usage, workmanship, lack of supervision, lack of knowledge, operation and cleaning were categorized under human factor.

References

1. Othuman Mydin, M. A., Ramli, M., & Awang, H. (2012). Factors of deterioration in building and the principles of repair. *Analele Universiti*, ISSN 1453-7397, 345–352.
2. Bakri, N. N. O., & Mydin, M. O. A. (2014). General building defects: Causes, symptoms and remedial work. *European Journal of Technology and Design, 3*(1), 4–17
3. Ramly, A. (2006). *Panduan Kerja-kerja Pemeriksaan Kecacatan Bangunan.* Hizi Print Sdn Bhd.
4. Ali, A. S. (2009). Cost decision making in building maintenance practice in Malaysia. *Journal of Facilities Management*, 298–306.
5. Ali, A. S., & Wen, K. H. (2012, September). Building defects: Possible solution for poor construction workmanship. *Journal of Building Performance, [S.l.], 2*(1), ISSN 2180-2106. Available at: http://spaj.ukm.my/jsb/index.php/jbp/article/view/20/14
6. Ahzahar, N., Karim, N., Hassan, S., & Eman, J. (2011). *A study of contribution factors to building failures and defects in construction industry. Procedia Engineering.* The 2nd international building control conference. 20, 249–255. Kuala Lumpur: Elsevier. https://doi.org/10.1016/j.proeng.2011.11.162
7. Ahluwalia, S. S. (2008). *A framework for efficient condition assessment of the building infrastructure.* University of Waterloo.
8. Chew, M. (2004). Maintainability of wet area of non residential building. *Journal Structural Survey*, 39–52.
9. Constantinos, A. B. (2005). Deterioration of European apartment buildings. *Energy and Buildings*, 515–527.
10. Wardhana, K. (2003). Study of recent building failures in the United States. *Journal of Performance of Constructed Facilities*, 151–158.
11. Abu-Tair, A. I., McParland, C., Lyness, J. F., Nadjai, A. *Predictive models of deterioration rates of concrete bridges using the factor method based on historic inspection data.* 9th International Conference on Durability of Building Materials and Components (DBMC), Brisbane.
12. Kian, P. (2001). A case study on total building performance evaluation of an intelligent office building in Singapore. *Civil Engineering Dimension, 3.*

13. Watt, D. S. (1999). *Building pathology: Principle & practice*. Blackwell Science.
14. Calder, D. (1997). Construction quality auditing. *Journal of Management in Engineering—J MANAGE ENG, 13*.https://doi.org/10.1061/(ASCE)0742-597X(1997)13:6(26)
15. Assaf, S., Al Hammad, A. M., & Al-Shihah, M. (1996). Effect of faulty desin and construction on building maintenance. *Journal of Performance Construction and Facilities*, 171–174.
16. Hassanain, M., Iftikhar, A., Al-Hammad, A.-M., Alshibani, A., Ibrahim, A., & Abdallah, A. (2020). POE of public higher education facilities in Saudi Arabia: Lessons learned from three case studies. *Property Management*. https://doi.org/10.1108/PM-06-2020-0037
17. BRE. (1991). Why do buildings crack? Digest 361, Building research establishment.
18. Richardson, B. A. (1991). *Defects and deterioration in buildings*. 9780419149200. https://books.google.com.my/books
19. Qazweeni, J., & Daoud, O. (1991). Concrete deterioration in a 20-year-old structure in Kuwait. *Cement and Concrete Research, 21*, 1155–1164. https://doi.org/10.1016/0008-8846(91)90076-T
20. Seeley, I. H. (1987). *Building maintenance* (2nd ed.). Macmillan Press Limited.
21. Ransom, W. H. (1981). *Building failures diagnosis and avoidance* (2nd ed.). E&FN Spon.
22. Chong, W. K., & Low, S. P. (2005). Assessment of defects at construction and occupancy stages. *Journal of Performance of Constructed Facilities*, 283–289.
23. Ahzahar, N., Karim, N. A., Hassan, S. H., & Eman, J. (2011). A study of contribution factors to building failures and defects in construction industry. *Procedia Engineering, 20*, 249–255.
24. Josephson, P. E., & Larsson, B. (2001). The Role of Early Detection of Human Error in Building projects. *CIB World Building Congress*, Wellington, New Zealand.
25. Choudhry, R. M., Gabriel, H. F., Khan, M. K., & Azhar, S. (2017). Cause of discrepancies between design & construction in the Pakistan construction industry. *Journal of Construction in Developing Countries, 22*(2), 1–18.
26. Chong, W. K., & Low, S. P. (2006). Assessment of Defects at Construction and Occupancy Stages. *Journal of Performance of Constructed Facilities*, 213–221.
27. Georgiou, J., Smith, J., & Peter E. D. (2014). A benchmark study on housing defects in Victoria, Australia. *International Journal of Construction Management*, 37–41
28. Salim, N. A., Salleh, N. M., & Zahari, N. F. (2016). *Design failure affecting maintenance management on public higher education institution in Malaysia*. International Building Control Conference, IBCC 2016 (pp. 00122, 1–7). Malaysia: EDP Sciences. https://doi.org/10.1051/matecconf/20166600122
29. Talib, R. (2014). Assessment of factors affecting building maintenance and defects of public buildings in Penang, Malaysia. *Journal Scientific Academic Publishing*, 48–53.
30. Othman, N. L, Jaafar, M., Wan-Harun, W. M., & Ibrahim, F. (2015). A case study on moisture problems and building defects. *Procedia-Social and Behavioral Sciences*, 27–36. Seoul, S.Korea.
31. Suffian, A. (2013). Some common maintenance problems and building defects: Our experiences. *Elsevier Procedia Engineering Science Direct*, 101–108
32. Mohd Sarman, A., Mohd Nawi, M. N., Che-Ani, A. I., & Mazlan, E. M. (2015). Concrete flat roof defects in equatorial climates. *International Journal of Applied Engineering Research, 10*, 7319–7324.
33. Ahmad, A. G. (2004). Understanding common building defects. *The dilapidation survey report, "Majalah Akitek", 16*(1), 19-21. First Quarter 2004.
34. Ali, A. S., Keong K. C., Zakaria, N., Zolkafli, U.,& Akashah, F.(2013), The effect of design on maintenance for school buildings in Penang, Malaysia. *Journal Structural Survey*, 194–201.
35. PWD. (2013). *Building condition inspection for existing building*. Kuala Lumpur: Malaysia Public Work Department.
36. NJ Green Building Manual. (2011). Rutgers Center for Green Building.
37. Ling, C. S., Ahmad, M. H., & Ossen, D. R. (2007). The effect of geometric shape and building orientation on minimising solar insolation on high-rise buildings in hot humid climate. *Journal of Construction in Developing Countries, 12*(1), 27–38.

38. Igor N. M., Thomas B., Oliver D., Shalini R, & Kenneth J. K., (2010). Enhancement of TiO2 visible light photoactivity through accumulation of defects during reduction–oxidation treatment. *Journal of Photochemistry and Photobiology A: Chemistry, 212*(2–3).
39. Ridout, B. (1999). *Timber decay in building*. Publisher : Taylor & Francis (December 29, 1999), ISBN-10 : 9780419188209, ISBN-13 : 978-0419188209.
40. Johar, S., Che-Ani, A. I., Tawil, N. M., Surat, M., S. N., & Kamaruzzaman, S. N. (2013). Preliminary survey and defects analysis of traditional timber mosques in Malaysia. *Wseas Transactions on Environment and Development, 9*(1), 12–23
41. Tan, C. K. W., & Corlett, R. T. (2012). Scavenging of dead invertebrates along an urbanisation gradient in Singapore. *Insect Conservation and Diversity, 5*, 138–145.
42. Pellitteri, P. J. (1995). *Controlling carpenter ants*. University of Wisconsin--Extension.
43. March, S., Blundell, J., & Ridout, B. (2000). Timber decay in buildings: The conservation approach to treatment. *APT Bulletin, 31*, 57. https://doi.org/10.2307/1504679
44. Che Ani, A. I, Sairi, A., Johar, S., Abd-Razak, M. Z, & Mohd Tawil, N. (2015). The practice of building inspection in malaysia total asset management. *Jurnal Teknologi*, 1–11
45. Ishak, N. H., Chohan, A. H., & Ramly, A. (2007). Implications of design deficiency on building maintenance at post-occupational stage. *Journal of Building Appraisal, 3*(2), 115–124.
46. Institute of medicine (US) committee on damp indoor spaces and health. Damp indoor spaces and health. Washington (DC): National Academies Press (US); 2004. Available from: https://www.ncbi.nlm.nih.gov/books/NBK215643/. https://doi.org/10.17226/11011
47. Almas, A.-J., Liso, K. R., Hygen, H. O., Oyen, C. F., & Thue, J. V. (2011). An approach to impact assessment of building in a changing climate. *Building Research & Information, 39*(3), 227–238.
48. ElSamadony, A., Hossny, O., ElHakeem, A., & Hussein, D. (2013). An asset management system for maintenance and repair of educational buildings. *International Journal of Scientific and Engineering Research, 4*(6), 2053–2064.
49. Haan, J. (2012). *Human error in structural engineering: The design of a human reliability assessment method for structral engineering.*
50. Samah, A. H., Tawil, N. M., Mahli, M., Che-Ani, A. I., & Abd-Razak, M. Z. (2014). Building condition assessment using condition survey protocol matrix: A case of school building. *Research Journal of Applied Science, 9*(9), 565–572.
51. Jorgensen, K. (2009). *Failure and defects in the building process—Applying the Bow-Tie approach*, CIB 2009
52. Horner, R. M., El-Haram, M. A., & Munns, A. (1997). Building maintenance strategy: A new management approach. *International Journal of Quality in Maintenance, 3*(4), 273–280.
53. Oladapo, A. A. (2006). A study of tenants' maintenance awareness, responsibility and satisfaction in institutional housing in Nigeria. *International Journal of Strategic Property Management, 10*(4), 217–231. https://doi.org/10.1080/1648715X.2006.9637554
54. Olanrewaju, A. L. (2009). Building maintenance management in Malaysia. *Journal of Building Appraisal*, 207–214
55. Olanrewaju, A. L., & Abdul-Aziz, A. R. (2014). *Building maintenance processes and practices: The case of a fast developing country*. Springer, Technology & Engineering, p. 331.
56. Bagdiya, N. V., & Wadalkar, S. (2015). Review paper on construction defects. *IOSR Journal of Mechanical and Civil Engineering (IOSR-JMCE)*, 88–91
57. Waziri, B. S. (2016). Design and construction defects influencing residential building maintenance in Nigeria. *Jordan Journal of Civil Engineering*, 313–323.
58. Okuntade, T. F. (2014). Effects of faulty design and construction on building maintenance. *International Journal of Technology Enhancements and Emerging Research*, 59–64.
59. Agwu, M. O. (2014). Perception survey of poor construction supervision and building failures in six major cities in Nigeria. *British Journal of Education, Society & Behavioural Science*, 456–472.

60. Glewwe, P., & Muralidharan, K. (2016). Improving education outcomes in developing countries: evidence, knowledge gaps, and policy implications. In S. Machin, L. Woessmann, & E. A. Hanushek (Eds.), *Handbook of the Economics of Education*, (pp. 653–743). (Handbook of the Economics of Education; Vol. 5). Elsevier. https://doi.org/10.1016/B978-0-444-63459-7.00010-5
61. Gordon, L. M., & Haasi, T. (1996). *Operation and maintenance in office buildings: Defining baseline.* aceee.org/proceeding/1996, (5.51–5.59).
62. Shanker, S. (2017). *Malaysian maintenance culture, or lack of it*. NST July 13, 2017.

Chapter 4
Building Condition Monitoring and Assessment

Abstract This chapter reviews method of assessment and inspection to gather the condition of the building. There are many different of methods used to evaluate the condition of element or components of the building. In this study, the BCA guideline of PWD were explained and is to be used in this study. The BCA result is then used as dependent variable for the study. Hence the method to be carried out for the study must be studied to fulfil the research objective which is to identify the condition of the building through condition survey. It is important that the outcomes of the assessment will produce consistency of the data collected. The method of research for data collection will be explained in the research methodology chapter.

4.1 Introduction

In principle the building should safe and protect the occupant from various weather condition, defects, failure, fire, being abandoned, especially for the building element and services in the building. So that, the building condition is very crucial to support decision making and to the management in achieving the service standards for maintenance. Building condition reflects the physical state of the buildings, and also its performance. So that the building should serviceable to suit the facilities in the building. As such, the organizations, buildings and maintenance managers must have knowledge in monitoring the condition of their buildings to prevent defects and failure of the buildings.

The damage and defects to the building will rise due to the inefficiencies of the system of building maintenance work. Defects and damage are common problems faced in any building without considering their age and types of construction. Besides that, without proper work system for handling those defects systematically, an effects and negative impact to the user of the building will generate. So that, to ensure the building safe and in long term quality, the various factors that caused the defects should be prevented. Several studies have revealed that building condition assessment is a process of evaluating a building condition systematically to counter the function needs of the building activities [1, 2].

S. Yacob et al., *Managing Building Deterioration*, Management in the Built Environment,
https://doi.org/10.1007/978-981-16-5860-0_4

According to Straub [3], condition assessment is a tool for evaluating and assessing the technical performance of the building to long-term maintenance expectations. Studies by DFES [4], shown that condition assessment is a tool to provide a systematic, uniform and objective basis for getting information on the state of the premises. Research by Kaiser [5], condition assessment is a process for inspecting and reporting the physical condition and also the functional performance of the building. The studies have identified there are benefits of knowing the current condition of the buildings [6] such as:

(i) Ability to plan and manage the service delivery of the building;
(ii) Avoidance of the building failures;
(iii) Preparation accurate projection of the future expenditure requirements;
(iv) Refinement of maintenance strategies;
(v) Risk management associated with asset failures.

According to INGENIUM and IPWEA [6], building assessment and monitoring has been subjective, using gut feel and local knowledge, including determining the expenditure of the operation and maintenance of the buildings. Ahluwalia [2], the current condition assessment is nevertheless highly subjective and its accuracy is highly dependent on the experiences and also lack of accuracy.

4.2 Building Condition Assessment Techniques

The building condition and its performance through its condition its self and its can be measured in different ways and methods. Building condition assessment performed through visual inspection by the assessor or inspectors or by an expert in specific building systems. The building system such as architectural work, civil, structure, mechanical, electrical and also building facilities based on the maintainability, security, functionality and sustainability.

Building condition assessment is growing especially in Malaysia. Hence, building condition assessment has been practiced in other countries and in particular in the United States, United Kingdom and many other European countries [7]. The literature suggests that it is necessary to assess the building condition systematically and comprehensively to maintain and to prolong the service life of the building hence to fulfill the satisfaction condition.

A building survey consists of an investigation and assessment of the construction and condition of building, which generally include the structure, fabric, finishes and ground [8]. The assessment of building condition is technically complex tasks, requiring knowledge, time and also equipment [9]. The author also noted that the building condition is assessed similarly in many countries although there are differences in assessment objectives. The assessment is on the basis of diagnosis of the extent of deterioration in the building elements. Research by Ramly [10]; suggests that building inspection is a testing, examination to a material or building product. Wordsworth [11], stated that the inspection is the process to assess the quality or service of building in order to achieve predetermined standards. The opinion of Wordsworth [11], the principle of building inspection at a building which must be done in accordance with the standards. An assessment of the condition of the building is the process of examining the entire building and infrastructure components. The assessment covered for internal and external work of building which including system, subsystems, components, subcomponents of the building.

4.3 Condition Assessment Systems

It has conclusively been shown by Ahluwalia [2] that since 1980s condition assessment tools have been developed for types of infrastructure assets. Table 4.1 shows examples of a condition assessment system developed for various types of infrastructure asset, such as pavement, railroad, bride, gas piping, sewer and also building [2]

A system of condition assessment developed to ease and facilitate the ranking of all the building components based on the amount needed rectifying work. The literature was studied to compare with the current system of condition assessment in present country. This research only focuses on existing buildings, hence the condition assessment for building are discussed and compared with government condition assessment.

Table 4.1 Condition assessment system for individual types of infrastructure asset

No	Condition assessment system	Type of infrastructure asset
1	PAVER	Pavement
2	RAILER	Railroad tracks
3	BRIDGER	Bridges
4	GRIPPER	Underground gas pipes
5	ROOFER	Roof
6	BUILDER	Building
7	RECAPP	Building
8	TOBUSS	Building

4.4 Building Condition Rating Scale

Building Condition assessment involved the systematic screening process to fulfil the long-term building maintenance expectations. The objective of condition rating is to determine the current condition of a building. A rating system also can be effectively used to predict future conditions. The condition indexes provide the values the condition of various components. The condition index scale and linguistic representation of building components usually varies from others countries, whether it can derive from the numeric values or alphabets. According to Pitt [12], it is important to clearly define the collected data to ensure the consistency and the reliability of the data. There are differences point scales presented in many countries to assess the building condition. For examples a six-point scale used according to the Dutch Standard for condition of the building without references to remedial work are as shown in Table 4.2.

Other research has shown that there is more range of scale used to assess the building condition Table 4.3 [2].

Meanwhile, Table 4.4 shows that a five-point scale were used to assess the condition of the existing building. Australia and United Kingdom used grade A to E to indicate the condition of their building. New Zealand, South Africa and United States used number 1 to 5 to indicate the condition of their building. Abbott et al.

Table 4.2 Six-point scale Dutch Standard for condition assessment of building

Condition rating	General condition description
1	Excellent
2	Good
3	Fair
4	Poor
5	Bad
6	Very Bad

Source Straub [3]

Table 4.3 Rating scales and representations

Asset type	Scale	Linguistic representation
Buildings	A-D	A-good, B-satisfactory, C-poor, D-bad
Buildings	1–8	1-excellent, 2-good, 3-adequate, 4-fair, 5-poor, 6-non operable, 7-urgent, 8-emergency
Any Asset	1–7	1-failed, 2-very poor, 3-poor, 4-fair, 5-good, 6-very good, 7-excellent
Buildings	1–4	1-no, 2-slight, 3-moderate, 4-severe
Buildings	1–4	1-good, 2-fair, 3-poor, 4-unsatisfactory

Table 4.4 Country comparison of the condition rating grade used for building

Government/private body	Condition grade (buildings)	Categories
Australia	A-excellent, B-good, C-adequate, D-poor, E-inadequate	Existing
New Zealand	1-very good, 2-good, 3-fair, 4-poor, 5-very poor	Existing
South Africa	1-excellent, 2-good, 3-average, 4-poor, 5-very poor	Existing
United Kingdom	A-good, B-fair, C-average, D-poor, E-bad	Existing
United States	1-very good, 2-good, 3-moderate, 4-poor, 5-very poor	Existing

[13], proved that five-point scale to be most effective scale compare to a three-point scale which is too coarse for reliable results, meanwhile more point rating scale such as seven-point is too fine and difficult for assessment staff to interpret consistently.

4.5 The Practice of Building Condition Assessment

Based on previous literature studies, there are different condition assessment methods executed in many countries. However, different results of various condition assessment methods are not a drawback in practice [3]. It is important to the organization to handle the condition assessment by their own method of assessment to ensure the reliability of the data collected. The following topic disscussed a few types of condition assessment standard used in other countries, to be compared with condition assessment standard used in Malaysia.

4.5.1 Dutch Building Condition Assessment Standard

According to Straub [3], the Dutch government building agency have standardised the condition assessment of building component and services. The standard application for property owners, managers and administrator, tenants, consultant, contractors and inspectors of control bodies. However, this standard is limited to condition assessment and not including maintenance priority action related to the building condition. It is reported that the standard was highly used by property manager and consultant in Netherlands to register the types of defects and the extent of these defects and also to record the status of building components (Vijverberg 2004).

4.5.1.1 Condition Assessment Process

The condition assessment processes as shown in Fig. 4.1.

Figure 4.1 explained that the first step executed is assessing the defects at the building components. The assessment based on the classification of intensity of

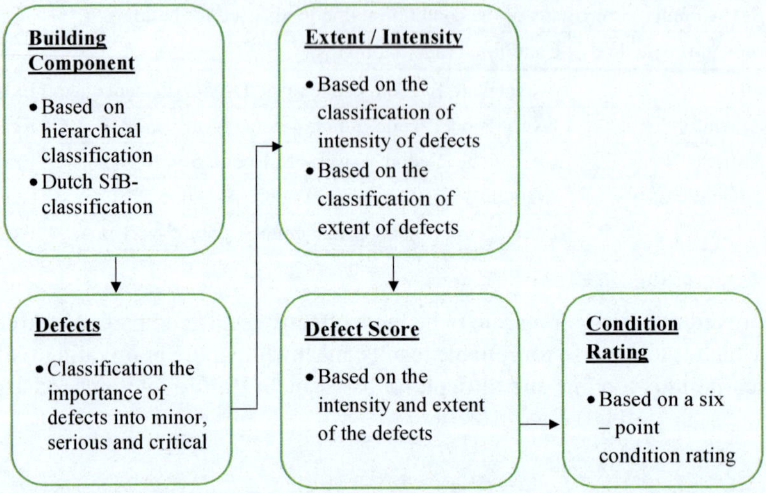

Fig. 4.1 Dutch condition assessment process. *Source* Straub [3]

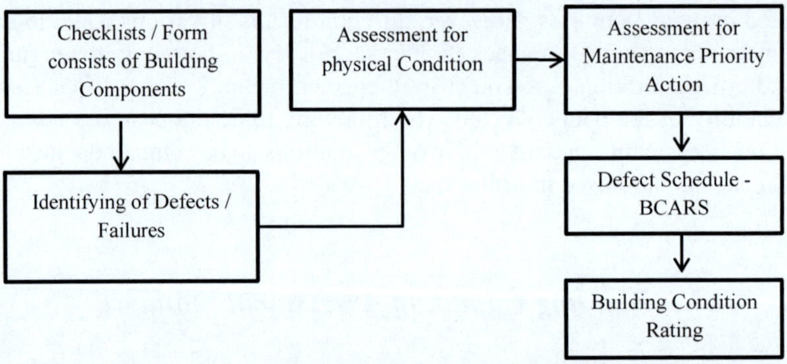

Fig. 4.2 PWD building condition assessment process. *Source* PWD JKR 21,602-0004-13 (2013)

defects as shown in Table 4.5. Meanwhile the extent of the defects explained in Table 4.6. Last step, the extent and the intensity of the defect combined to come up with the condition rating of the building. The six-point scale of condition rating shown in Table 4.2 as explained before.

Table 4.5 Classification of intensity of defects

Intensity class	Name	Description
Intensity 1	Low	The defects is hardly visible
Intensity 2	Middle	The defects is progressing
Intensity 3	High	The defects cannot progress any further

Source Straub [3]

Table 4.6 Classification of extent of defects

Extent class	Percentage	Description
Extent 1	<2	The defects occurs immediately
Extent 2	2–10	The defects occurs locally
Extent 3	10–30	The defects occurs regularly
Extent 4	30–70	The defects occurs frequently
Extent 5	≥70	The defects occurs generally

Source Straub [3]

As conclusion, the standard for Dutch condition assessment is limited to large scale property and not meant for preparing the annual maintenance budget and planning work [3].

4.5.2 *Queensland Building Condition Assessment Standard*

The Queensland building condition assessment standard was published to be used in 2004 under Department of Housing and Public Works. The standard was developed to evaluate the physical state of building elements and services and to assess the maintenance needs of the facility. The actual condition of building was assessed for its individual elements and services, in comparison to asset owners' specified condition standard. The maintenance work then identify and rank according to the priority to maintain specified condition standard.

4.5.2.1 Condition Assessment Process

Table 4.7 explains the five steps of the condition assessment process which can assist the department to refine the existing maintenace planning procedures.

The first step in the assessment, the standard for overall building to be identified with a rating S1 to S5 with functional purpose. Table 4.8 shows the condition standards to be determined by the appropriate standard required at facility or individual building level.

Secondly, the condition assessment involved in the evaluation of the physical condition of building. The rating used in this standard as shown in Table 4.9. Five-point scale between 1 and 5 was used in this standard to rate the extent of the defects. The rating 5 indicates excellent condition shows that the building have no defects and are as new condition. The rating 4 inidicates good condition with minor defects, superficial wear and tear, some deterioration to finishes and no major maintenance required. Meanwhile the rating 3 shows the building is in fair condition. The building is in average condition, significant defects are evident, worn finishes require maintenance, services are functional but need attention and deferred maintenance work

Table 4.7 Stages of the condition assessment process

Stage	Description
Scoping of services	• Determining the requirement for information and other data to meet asset management obligations • Developing service specifications
Procurement of services	• Procuring the condition assessment service from a service provider using appropriate arrangement
Planning the implementation	• Planning the implementation of condition assessments based on the requirements established through the scoping stage
Implementation	• Reviewing data from previous assessment, removing references to completed or redundant works, and making corrections to ensure that appropriate areas are targeted during condition assessments • Undertaking the assessment in accordance with the agreed scope and timeframes • Recording the data from the assessments and updating any existing data from previous assessments
Reporting	

Source Queensland Department of Housing and Public Works (2017)

Table 4.8 Queensland condition standards

Functional purpose	Specified standard	Rating
Highly sensitive purpose with critical results or high profile public building	Building to be in the best possible condition. Only minimal deterioration will be allowed	S5
Good public presentation and a high quality working environment are necessary	Building to be in good condition operationally and aesthetically, benchmarked against industry standards for that class of asset	S4
Functionally-focused building	Building to be in reasonable condition, fully meeting operational requirements	S3
Ancillary functions only with no critical operational role	Building to meet minimum operational requirements only	S2
Building is no longer operational	Building can be allowed to deteriorate, however, must be marginally maintained to meet minimum statutory requirements	S1

Source Queensland Department of Housing and Public Works (2017)

exists. The rating 2 indicates the building in poor condition, badly deteriorated, potential structural problems, major defects and component fail frequently. The rating 1 indicated the building in very poor condition.

Thirdly, the ranking scale of maintenance priority should be assessed to provide an indication of recommended maintenance work. Table 4.10 shows the condition assessment ranking scale used in this standard.

Table 4.9 Condition Index

Rating	Status	Definition of rating/condition of building asset
5	Excellent	• No defects • As new condition and appearance
4	Good	• Minor defects • Superficial wear and tear • Some deterioration to finishes • Major maintenance not required
3	Fair	• Average condition • Significant defects are evident • Worn finishes require maintenance • Services are functional but need attention • Deferred maintenance work exist
2	Poor	• Badly deteriorated • Potential structural problems • Inferior appearance • Major defects • Components fail frequently
1	Very poor	• Building has failed • Not operational • Not viable • Unfit for occupancy or normal use • Environmental/contamination/pollution issues exist

Source Queensland Department of Housing and Public Works (2017)

Table 4.10 Condition assessment ranking scale

Priority ranking	Definition
1	Works needed to: • Meet maintenance related statutory obligation and due diligence requirements • Ensure the health and safety of building occupants and users • Prevent serious disruption of building sctivities and/or may incur higher costs if not addressed within 1 year
2	Works that: • Affect the operational capacity of the building • Are likely to lead to serious deterioration and therefore higher future repair costs if not addressed between 1 and 2 years
3	Works that: • Have minimal effect on the operational capacity of the building but are desireable to maintain the quality of the workplace • Are likely to require rectification within 3 years
4	Works that: • Can be safely and economically deferred beyond 3 years and ressessed at future date

Source Queensland Department of Housing and Public Works (2017)

The results from the condition assessment should be recorded in report. The condition assessment report provided the necessary information for the development of a condition-based maintenance program.

4.5.2.2 Royal Institute Chartered Surveyor (RICS) Condition Assessment Standard

RICS standard is a tool that are used by surveyor or building inspector to carry out the inspection to ensure the reliability of the data collected. The inspection must performed by surveyor who has sufficient knowledge and competence comply with the requirement in this standard especially about the defects, elements and remedial works [3]. To assess the condition of the building, standard list of defects, condition parameter and condition rating should be clearly identified by the surveyor or inspectors.

4.5.2.3 Condition Assessment Process

The methodology as shows in Table 4.11 should be used by surveyor to establish the condition rating for each element or sub element.

Table 4.11 Methodology to establish the condition rating

Steps	Description
1	Identifying the elements and the sub-elements. The surveyor inspect the building by identifying all the elements and makes notes on the condition in the site notes
2	Condition rating the elements and sub elements. The surveyor should appy a condition rating to all the elements and sub elements as shown in Table 4.12
3	Establishing the element and sub-element rating. The condition rating to be shown in the elemental condition rating for each element must be the worst one identified by this process
4	Report writing. The report writing prepared to provide the judgment of the condition rating

Source RICS Building Survey (2016)

Table 4.12 RICS condition ratings

Rating	Description
3	Defects that are serious and/or need to be repaired, replaced or investigated urgently
2	Defects that nedd repairing or replacing but are not considered to be either serious or urgent. The property must be maintained in the normal way
1	No repair is currently needed. The property must be maintained in the normal way

Source RICS Building Survey (2016)

4.5.2.4 Condition Rating

The condition rating are shown in Table 4.12. The table shows that rating 3 must be used where the defect is of serious nature and in need for immediate repairs or where the surveyor feels unable to reach the necessary conclusion with reasonable confidence. The rating 2 must be used where defects that needs repairing or replacing but the defect is not considered to be serious or urgent. Meanwhile the rating 1 is used when there are no indications of present or suspected defects that require the undertakin of specific repair.

To conclude, the RICS standard is to carry out the condition assessment developed to ease the surveyor. It is essential that the condition assessment is performed by a surveyor who has sufficient knowledge and competence necessary to carry out a detailed condition assessment of the building. The use of the standard is also to ensure the client's requirements are identified besides to ensure thorough detailed building inspection will carried out.

4.5.3 The Current Practices Condition Assessment in Malaysia

This study focuses more on the tools used to assess the building conditions in Malaysia. The condition assessment guidelines and systems such as Building Condition Maintenance Assessment System (BCMAS) and Building Condition Assessment Rating Systems (BCARS) will be discussed.

(a) BCMAS was developed by Kumpulan Ikram Sdn. Bhd (KISB) for Public Work Department (PWD) Malaysia to carry out inspections of Federal Government's building facilities that have passed Defects Liability Period (DLP) throughout Peninsular of Malaysia. The objectives of the development of BCMAS are:

    (i) To ascertain the conditions of the relevant Federal Government's building facilities by carrying out inspections on the condition of the building facilities and existing building facilities maintenance practices
    (ii) To develop suitable BCMAS including manual, software and check-lists that are appropriate and can be integrated with other maintenance systems
    (iii) To enhance the competency of the personel involved in the effective implementation of inspection and maintenance works of Federal Government's building facilities.

(b) BCARS was developed by PWD Malaysia to improve the Government service delivery toward effective decision making for building maintenance. BCARS is developed to assess and assist the ranking of all components of the buildings. The objectives of the development of the BCARS are:

    (i) Determine the condition of the building;

(ii) Determine the overall rating as final result and it will easily access and understood;
(iii) Apply to all types of building covered by block, level and the use of the building;
(iv) Deliver the finding of assessment consistently and correctly identifies the level of maintenance;
(v) Be applied by building inspector or assessor who receives a training course;
(vi) Be accepting and approve to be implemented to all government buildings.

4.6 Building Condition Maintenance Assessment System (BCMAS)

A BCMAS application system developed by PWD Malaysia to adopt an asset management system to the Federal Government's building facility that is total and systematic. One of the aspects of a good asset management is getting a condition assessment on building facilities to the existing Federal Government so that all defects/damage are identified and repaired to ensure safety and functionality of the relevant building facility.

The Federal Government of Malaysia is the owner of many buildings that house their executive, administrative and operation facilities. Many of these buildings are not systematically inspected and maintained (PWD Contract Document, JKR/IP/BSFB/60/2010). Inspection, maintenance and repairs are normally carried out on need or ad-hoc basis. The current system is more reactive rather than preventive in nature. As highlighted by Ahluwalia [2] there are lack integration between maintenance/repair and capital renewal functions, the need for efficient inspection, and need for less-subjective condition assessment.

This was also highlighted by Ismail et al. [14] who stated the maintenance management in Malaysia is still reactive and most of the organizations implementing the conventional method to manage the buildings. Hence, proper and adequate tools are necessary to enhance the inspection and maintenance work such as comprehensive inspection, manual, software and checklist. Previous studies showed that it is important to have a systematic condition assessment or inspection system towards comprehensive maintenances and sustainable asset management [13].

Until 2008, the Government of Malaysia has no standard form of inspection of buildings/government assets to be implemented. BCMAS application system development is intended as a guideline so that the PWD agencies located throughout Malaysia are to make an inspection and reporting of inspection in one On-line application system as a One-Stop Centre. Reports issued can justify the purpose of maintenance of the buildings inspected. The project was created in accordance with the decision of the instructions of the Federal Government of Malaysia in 2007 and general orders, Chapter E, clause V. "The responsibility of the Department of public works".

4.6.1 *Scope of Work*

BCMAS will be used to determine the functionality and safety of buildings and services in the buildings on "as-is" basis. "As-is" basis is dependable on the best information available at the time of the inspection. As highlighted by Building Code Queensland [15], the effectiveness of an inspection is enhanced by the availability and the clarity of the information gathered and also to the compliance assessment according to the guideline. This includes defects, damages or deteriorations which can be obviously ascertained by the expert. The scope of work under BCMAS is as presented below:

(i) Carry out inspection for Federal Government's building facilities;
(ii) Inspect randomly on a sample basis damages which are caused by normal 'consumable usage';
(iii) Provide adequate and suitable expertise, resources and equipment for inspection;
(iv) Preparation of checklist, Quality Plan and Inspection Methodology;
(v) Come out with the recommendations and cost estimates to rectify defects, damages or deteriorated conditions;
(vi) Preparation of reports and recommendations; and
(vii) Come out with analysis and assessment of all data and give the overall comments and recommendations

4.6.2 *Methodology*

4.6.2.1 Desk Data Collection

The data collection involved the compilation of all necessary and relevant information about the buildings. The data collection carried out are based on the three (3) sub-categories:

(i) Obtained the numbers or list of the related buildings. General information such as the name of the project, site location, acreage, terrain, and usage;
(ii) Data collection for building inventory at each project site, such as age, usage, numbers of floor, built up areas, conditions and special design/features;
(iii) Detailed information for buildings such as as-built drawing for architectural, civil, structural, mechanical and electrical. Another information is operation and maintenance manuals and records.

4.6.2.2 Logistic

The numbers of the buildings to be inspected are divided into four (4) zones. The Central Zone which covered Perak, Selangor, Negeri Sembilan and Pahang. The Southern Zone covered Johor and Melaka. Meanwhile, the Northern Zone covered Perlis, Kedah and Pulau Pinang. Eastern zones covered Kelantan and Terengganu.

The inspection will be carried out by an inspector who will be nominated under four (4) regional offices for each zone selected.

4.6.2.3 Checklists

The checklist for BCMAS was prepared to cover the following building facility components:

(i) Architectural which included internal and external finishes such as cladding, tiles, ceiling, and roofing. Others are windows, doors, handrails, fittings and built ins, decorative elements, sanitary fittings and accessories, rainwater gutters and down pipe and other related works;
(ii) Geotechnical which covered natural, fill and cut slopes;
(iii) Civil which included road and drainage, water supply work, sewerage works, retaining walls and others;
(iv) Structural which are walls, beams, slabs, columns, roof structures / trusses, and water proofing;
(v) Mechanical which included air conditioning and mechanical ventilation, fire-fighting system, lift and escalator, internal hot and cold water plumbing system, internal sanitary systems, boiler, water supply pumping system and any other fixed mechanical systems in the buildings and around the compound; and
(vi) Electrical which consisted of low voltage (LV) and high voltage (HV) electrical systems, generator set, communication system, public address and sound system, building automation system, data networking cabling system, security and others related work.

The inspection checklist included checking for compliance of relevant statutory and regulatory requirements with respect to operation and maintenance aspects.

4.6.2.4 Data Analysis

Data available for BCMAS were analysed to generate useful information and to assess hazard risks such as the information gathered during the initial phase, building size, complexity, and the number of floors, terrain, age and usage. The building condition of the facilities evaluated for remedial actions and recommendation together with the cost estimation were also analysed. The data analysis will be carried out in the following three (3) stages:

(i) Building inventory statistic: age of building, number of floors, usage of buildings, built up areas, terrain, availability of maintenance records, as-built plans and operation and maintenance manuals.
(ii) Preparation of Inspection Report
(iii) Analysis of field data and recommendations.

4.7 Building Condition Assessment Rating Systems (BCARS)

Public Works Department (PWD) of Malaysia in year 2008 have developed, expanded and implemented the Asset Management to the organization to improve the government service delivery. Integrated and comprehensive Asset Management is to ensure that the asset are in a good condition, secure, economical, and longer life cycle as well as to provide quality services to the people [16]. Mihai et al. [17], noted that asset management as the function and maintaining buildings accordance standard which include benefit outcomes to stakeholder requirements using systematic tools. This covered for asset registers, asset condition monitoring, deterioration prediction, valuation of asset and others.

PWD has been developed BCARS in order to improve the current process of building inspection. The systems are more standardized to respond the challenges in managing and maintaining the assets. BCARS have been developed according to the requirement of Government Total Asset Management Manual [18] together with [19] (JKR 21602-0004-13). Abbott et al. [13], through their research reported that BCA is to facilitate the degree of the component reflected with the amount needed repair. The data collected will produce the consistency, relevant and useful information. The objective of the assessment should be clearly understood to allow the assessors to take precaution action within limited time of inspection. Maintenance can be effectively planned and managed if the priority maintenance action is properly quantified.

Building inspection is very important to assess the performance, safety and comfortable condition to achieve the organization goals in facilities, accommodation high quality work in an environment. Operating cost will increased with low level of building performance and this situation can give negative impact to the organization.

4.7.1 Standardization

Building assessment at a building must be carry out in accordance with the standards and related guideline. Wordsworth [11], mentioned that the condition of the building is a process of diagnosis the whole entire building depends on component inspection criteria developed based feedback from the user base. This included infrastructure components, including equipment for mechanical and electrical systems, building frames, internal structure and finishes as well as building sites.

Building condition survey is the method which depend on the criteria and objective of assessment. The result of the assessment will be useful for examining and improving the quality of design. The structure, architecture, mechanical, electrical, outdoor work and building facilities are the complete set of components for building inspection made valuation based on the maintainability, security, functionality and sustainability. The role and scope of the inspection should be understood by the examiners comprising Building Surveyors and other profesional such as Engineer or Architect.

According to Straub [3], the different results of various condition assessment methods are not a drawback in practice. So that it is very crucial that the assessor or inspector have same way in terms of knowledge and skill to assess or evaluate the defects. Kempton et al. [20], noted that the practice of condition assessment yields variable results due to subjective perceptions of the different inspectors. Hence, it is important the needs of training to the assessor to ensure the reliability data collected by the assessors.

The PWD Building Condition Inspection for Existing Building is a guide to the building assessor to do the inspection work at site besides as a references for consistency of data collected and report preparation. The purpose of the guideline is to ensure the assessor comply with inspection prescribed guidelines hence to provide the related and reliable rating and information about the defects assessed. The standard is not limited to condition assessment only but also to the maintenance planning and priotizing maintenance work. Some processes and regulations have to be complied with by the parties involved so as to ensure the implementation of the inspection are carried out in a more systematic manner.

The standard is used to fulfil the objective of the inspection which applicable for the following stages:

(i) For a purpose of asset registration;
(ii) For a purpose of operating and maintenance;
(iii) During the operation and maintenance of the building;
(iv) For a purpose of rehabilitation, renovation and upgrade;
(v) For a purpose of conservation of buildings;
(vi) For a purposes of risk assessment and safety;
(vii) For a purpose inspection after a disaster.

4.7.2 Development of BCARS

The study on the existing methods of assessment BCMAS 2010 were analysed. The purpose of the BCMAS is to expedite maintenance works, ensure proper financial management, assisted in planning and monitoring. According to supporting data (BCMAS report 2010), the method were not accurate enough and does not achieve the objectives. The development of the new standard is based on the requirements towards

National Assets Management. International rating comparison, namely Australia, New Zealand, South Africa, United Kingdom and the United States are reviewed for reasonableness of the use of this identification of BCA in Malaysia. The results of the BCA are a more systematic approach in building maintenance practices.

Guidelines for inspection of Building Conditions was developed to provide guidance on the methodology of implementation of inspection and the assessment of the building, which is simple enough to be used by the building inspectors or officers as stated in the general orders chapter 27 Clause E stating that the PWD is responsible for examining government buildings every year.

These guidelines also are to assist the inspectors or assessor in order to carry out inspections at the sites which is simple, consistent, systematic and holistic. The development involved with the preparation of the following documentations:

(a) Guidelines for Inspection of the Physical Condition of the Building;
(b) Building and physical condition Inspection Checklist;
(c) Building Condition Assessmet Rating System (BCARS).

The rating results can be accepted depends on the frequency of maintenance. The rating results also can be used for decision making during planning of maintenance work. Table 4.13 shows the expected outcomes of the BCARS development by PWD which projected can solve the defects issue in the buildings either for long term or the short, preparing an estimate of repair costs or replacement and maintenance budget more accurately.

4.7.3 Piloting

Various session workshops were carried out to discuss with the expert, experience assessors, engineers and also architects during the development phase of the standard guideline of BCA. The development and rating contributed to one of the requirements to the Building Performance Assessment (BPA) as a whole as the requirements in Standard Procedure for un-moveable of Government Asset [21]. The assessment methods were then presented in order to discuss, to study and to give comment for the improvement before conducting the pilot project. A pilot projects involved 15 types of government buildings with functions as office, health clinic, quarters and learning institution. The training course was attended by more than 300 numbers of officers from multi discipline such as building surveyors and engineers since 2011 to gather the data. The purpose of the training course is to ensure the reliability and validity scores of the rating given and the use of standard procedure appropriately.

Table 4.13 The expected outcomes of BCARS development

Results of the BCA	The outcomes of BCA	Solutions to overcome problems
Standard rating conditions/conditions of buildings	(a) Produce a data access system information and situation/condition the building easily and quickly; (b) Ability assets serves as the level of services required; and (c) Able to produce deterioration curve for buildings	(a) Inefficient maintenance work (b) Less supervision and monitoring (c) The issue of disability and building failures (d) Activities and consumer attitudes
2. Maintenance work scheduled 3. Cost estimates	(a) Program maintenance work condition-based maintenance (b) The estimated cost of short term and long term	(a) System maintenance work which is not planned and systematic (ad-hoc) (b) Inadequate maintenance budget (c) Less supervision and monitoring
4. Maintenance needs and long terms costs	Predictions of budget needs for repairs, upgrades in the future	(a) System maintenance work which is not planned and systematic (ad-hoc) (b) Maintenance requirements in the design of buildings (c) The issue of disability and building failures (d) Less supervision and monitoring
5. Solid/schedule waste disposal estimation	Comply with solid and hazardous waste management ACT 672 and Environmental Acts (Environmental Quality Act) EQA 1974	(a) Waste management plan for existing buildings

Source PWD Brainstorming Workshop (2009)

4.7.4 Improvement and Approval

A presentation session and meeting were held to get the approval to use the guideline. The purpose of the presentation is to inform and to request the consent of the Board to approve and adopt the guidelines for the examination and assessment of the condition of Existing Buildings and building inspection checklist to be used. This guideline was developed from the string which comes from the revenue issue and teaching and learning building acceptance inspection and building inspection reports and other related reports. Rating maintenance BCA is validity within one (1) from the date of certificate issue. Each rating should be made the following year. The standard was approved to be implemented in 2013 to the government building in Malaysia. The strategic plan was prepared to the implementation of the BCA and involved a series of course, training the trainers, seminars which is from various ministry since the TPATA was launched in year 2012.

4.7.5 *Method of Assessment*

The overall process of the assessment according to the standard as follows:

(a) Appointment of team of inspection which consist of multidisciplinary professional;
(b) The execution of preliminary inspection to gather related information and current condition of the building;
(c) The preparation of inspection planning program;
(d) Preparation of the related checklist according to the current condition of the building;
(e) Appointment with expert especially for structure, mechanical or electrical issue which involved with safety and security of the component or systems;
(f) The execution of the inspection work.

BCA will be carried out according to the PWD Standard. The inspection is based on the visual inspection by using appropriate tools and equipment. If necessary, involving sound, heat, smell, light and others related information. The assessment physical involved observation at site visit. Inspections done are subject to the scope of limitations [22]. This is the initial screening examination for the purpose of evaluating the condition of the building by Building Inspector. The result analysis of inspection shall be referred to experts for follow-up action. Figure 4.1 shows the condition assessment process according to the guideline.

4.7.5.1 Inspection Checklist

A checklist was developed according to the standard. The purpose of the checklist is to register all the related information of the building. This is include the identification of the building, characterisation, types, time, date, name of inspector and location or space of the building. A checklist used to evaluate the condition of the defects and maintenance priority action taken according to the visual inspection carried out by the inspector or assessors. Asset condition checklist was useful to know the status of the output and as input of the operation and maintenance of the building, and also to make a recommendations to the experts for follow-up action. The data collected then complied for updating and transferring to the defects sheets and BCARS for analysis.

The standard checklist was divided into a few parts which is which is building informations forms, civil and structure, mechanical, electrical and external work forms. The checklist is developed for the following purposes:

(i) To be used during inspection work accordingly;
(ii) To ensure the consistency of the data collected;
(iii) To assist in the analysis and preparation for comprehensive report.

Building informations parts includes building inspection services, building and information and reference documents required for inspection. The building measured should be classified by the category and the criticality of the building. This is important to record the characteristic of the building. The characterisation of building such as the category of building useful for record and for statistical analysis [9].

For civil and structure part consists of the elements of architects and civil law comprising basic system, structure, exterior parts, roof, interior building, staircase inside, ceiling, fixtures and also construction. For mechanical part consists of the mechanical systems covering piping system, air-conditioning system, distribution system, fire-fighting systems, elevator and elevator as well. Electrical parts consist of electrical systems that include lighting systems, communications and also security. Meanwhile for external work part consists of the system boundaries, roads, water reticulation, drainage, retaining walls, sewage treatment plants and soft and hard landscaping. The standard developed according to the building element hierarchy with three levels of details such as systems, subsystems and components. The overall checklist which covered for all scope of work was attached in Appendix A.

The infrastructure facility is complex structure and consists of many components and subcomponents, including section, systems, assemblies, and elements. So that, asset or building hierarchy is important to ease the data collected and recorded as useful information for decisions making. This also supported by Mohseni [23], condition assessment is done at detailed level such as components or component-type level of buildings to ease the process of assessment. Due to this situation, the assessor needs a clearly defined and hierarchical classification of building components [3]. Abbott et al. [13] through his study, highlighted condition of building are given for each block and building stories.

4.7.5.2 Identifying the Defects

Identifying the defects is important to assess the extent of defects associated with the condition of the building. At the first step, it is important to identify the types of defects rise in the building in order to identify the factors associated with those defects.

This statement was supported by Straub [3] stating that the assessing of defects occurs first and without the information one could not formulate maintenance activities and estimate the costs. Subsequently the inspector passes through the following condition parameter, importance and extent of defects [3]. The intensity of the defects will affect the building component or element to function properly. This statement also supported by Straub [3], the condition of the building influences by the extent of defects which affect the overall condition of the building. All the findings are assessed and recorded in building condition schedule and in defects sheets in photographs capture. The finding of the defects to be filled and checked based on the extent of the defects. The priority defects or damage needs to be given priority.

4.7.5.3 Physical Conditions Assessment

Condition assessments are conducted at every level of element or component by assessing the physical condition of the building associated with the maintenance actions. Assessment to each defect must be assigned according to the standard by assessor or building inspector. The defects are assessed by evaluate the current condition or performance with the original state of conditions. The condition of the buildings based on the following criteria such as functionality, security, maintainability and sustainability and summarized in the preparation of reports.

This is very important and critical to the inspector to ensure that the defects assessed accordingly without influence to the overall rating of the building. Table 4.14 shows that the defect is assessed on five-point scale. Grade 1 is a very good condition with no defects, grade 2 is good condition with minor defects, grade 3 is fair condition with major defects, grade 4 is poor condition with major or minor defect, critical, not functioning as agreed service level and grade 5 is very poor condition with very critical, not functioning, risky to safety and health.

According to Ramly [10] there are various guides to take into consideration when assessing the defects such as current state of the component or element, the size and the extent of defects, total cost and time to rectify the defects and the total number of defects. Previous research proved that a five-point colour-coded rating system has been proved to be most effective [13]. This will ease the method of assessment to interpret consistently, making reports more user-friendly, save time, comprehensive and also accessible to non-technical users to use the information.

Table 4.14 BCA physical condition level

Grade	Assessment scale	Description
1	Very good	As new, no defect, performing as intended
2	Good	Minor defect, good condition, performing as intended
3	Fair	Major defect, moderate condition, still can functioning with supervision
4	Poor	Major or minor defect, critical, not functioning as agreed service level
5	Very poor	Major or minor defect, critical, not functioning as agreed service level. Very critical, not functioning, risky to safety and health

Source PWD JKR 21602-0004-13, (2013)

Table 4.15 Maintenance priority actions

Priority	Scale	Description
Normal	1	No defect or damages, element/component well maintained
Routine	2	Minor of defects/damages, needs for monitoring, repairs, replaced to prevent serious defect/damages
Repairs	3	Major defects/damages, needs for major repairs and replacement
Rehabilitation	4	Critical/serious defects/damages, needs for urgent and immediate repairs
Replacement	5	Critical/serious defects/damages, needs for urgent replacement, refer to expert detail inspection/Expert judgement

Source PWD JKR 21602-0004-13 (2013)

4.7.5.4 Maintenance Priority Assessment

Table 4.15 shows that in BCA the physical condition assessment have to link with maintenance priority action to be taken prior to defects findings. The maintenance actions taken will based to priority. Scale 1 with the condition with no defects with normal priority, scale 2 is routine priority with minor defects, scale 3 is repairs priority with major defects, scale 4 is rehabilitation priority with critical or serious defects which needs for urgent and immediate repairs and scale 5 is replacement priority with critical or serious defects which needs for urgent replacement. This situation needs to refer to expert detail inspection or expert judgement.

Abbott et al. [13], stated that the condition assessment rating is linked to the related maintenance actions and types. This is due to the different types of component with different conditions may require the range of maintenance actions such as preventive, condition-based maintenance, rehabilitation and replacement. Hence, the importance of maintenance planning is prioritising maintenance, repairs, rehabilitation and replacement works dealing with the risk of the failures within available budgets. The allocation of cost will be arranged based on the priority according to the percentages of the defects found. Action taken based on maintenance priority is important to prepare the planned maintenance budget [10].

4.7.5.5 Defects Score and Building Rating

Research conducted by Mohseni [23] has come out with new assessment method for condition and rating of the building. She also stated that the earliest contribution to building rating was based on rating criteria which can applied to any type of building [8]. Mohseni [23], also carried out the inspection by adopting the Navil Matrix as a method assessment and analysis. BCARS is a defect score which indicate the combination of the extent of defect and the maintenance action. This is the same method of calculation proposed by Mohseni [23] where a rating system was used to predict future condition. Table 4.16 shows the matrix analysis score defect parameter for building condition and maintenance action.

The analysed defects data are summarised and sorted for rating and analysis purposes. The major defective elements were identified to give the priority of action taken. The score obtained from the building condition schedule determine the level of defects finding such as very good, good, fair, poor and very poor. The results were analysed and delivered to the conclusion by findings of final categorization of building ratings. One component or element may show more than one defect. The building components may show more than one defects [3]. The overall results of building rating derived from the total mark from sum up matrix analysis divided by total number of defects. Table 4.17 shows the overall building rating to be delivered to the main conclusion. The five-point colour-coded building rating is shown as a final result of the assessment.

Table 4.16 Matrix Analysis building physical condition and maintenance priority actions

Scale		Maintenance priority action				
		5	4	3	2	1
Building physical condition	5	25	20	15	10	5
	4	20	16	12	8	4
	3	15	12	9	6	3
	2	10	8	6	4	2
	1	5	4	3	2	1

Source PWD JKR 21602-0004-13 (2013)

Table 4.17 Overall condition rating

Rating	Physical condition	Action matrix	Score
A	Very good	Preventive maintenance	1–5
B	Good	Condition based maintenance	6–10
C	Fair	Repairs	11–15
D	Poor	Rehabilitation	16–20
E	Very poor	Replacement	21–25

Source PWD JKR 21602-0004-13, (2013)

Table 4.18 Number and generic for BCMAS defect

No	Name	BCMAS	
		Element	Generic criteria for defect
1	Architectural	29	27
2	Civil and structure	32	30
3	Electrical	41	39
4	Mechanical	53	38

Source BCMAS

4.8 Comparison Between BCMAS and BCARS

4.8.1 Building Hierarchy Information

Condition Assessment is defered according to the guideline developed based on the objectives and requirement of the inspection and clients. BCMAS was developed and implemented in 2010. The use of BCMAS checklist can be divided into three (3) phases which is before, during and after inspection. This covered for element and type of defects found during the inspection. The number and generic for BCMAS defect are as shown in Table 4.18.

BCARS was developed in 2012 and first launced by the Government and used in 2013. The components or elements inspected was developed based on building hierarchy which is system, subsystem and components [24]. In a different study, Mohseni [23] examined that making a decision depend on detail in a building hierarchy. Beside that according to INGENIUM [7], building condition monitoring based on a building hierarchy with three (3) levels such as component, element and sub elements.

Ahluwalia [2], stated that the benefit of the building hierarchy are to facilitate the process of revising components, to evaluate the performance of the building in order to ensure the safety of the components within sufficient cost of maintenance or repair works. The study by Ahluwalia [2] also identified five (5) main elemental classification systems used for data effectively such as the American UNIFORMAT classification, Canadian CIQS classification, RICS classification, UNICLASS and CEEC classification. Table 4.19 is a summary of BCARS representations of building information and possible defects found during the inspection.

4.8.2 Rating Scale, Representation and Inspection

According to Mohseni [23], a condition assessment is the method of classification of physical deterioration for condition of the buildings. For example, he also found that most commonly condition rating system used for building is the basic 1 to 5

Table 4.19 BCARS representions of building information and generic for BCARS Defect

No	Name	BCA—BCARS			
		System	Subsystem	Component	Generic criteria for defect
1	Architectural and civil structure	9	29	116	38
2	Mechanical	6	24	96	26
3	Electrical	4	14	62	26
4	External Work	19	43	129	38

Source PWD JKR 21602-0004-13 (2013)

rating. Some organzition have more point scale of condition ratings such as 1 to 10, 0 to 10, 0 to 100. He also concluded that all the assessment are subjective and there is no absolute right answer for condition rating scales and it is appropriate to use whichever rating system that works for the particular organization.

Table 4.20 shows the condition rating scale for BCMAS approach. The condition 1–3 were used by representing condition 1 as good condition with no defect and performing as intended. Condition 2 is unsatisfactory condition, the building performing whilst exhibiting defects or deterioration which requires repairs. Meanwhile condition 3 is critical condition where the components or elements exhibiting defects or deterioration, not operating as intended which needs urgent attention or repair.

Table 4.21 shows the rating scale 1–5 used during inspection for BCARS. Scale or grade 1 is the component or element in a very good condition, as new with no defects and performing as intended. Condition scale 2 is good condition with minor defects, component or element performing as intended. Condition 3 is fair condition with major defects, component or element still can functioning with supervision. Meanwhile condition 4 is poor condition. Component can be major or minor defects, critical, not functioning as agreed service level.

This also depend to the function and the suitable of the space or room design. Condition 5 is very poor condition and very critical, component and element not functioning, also risky to safty and health. The scale used in BCARS are more comprehensive in order to make the data or the findings of inspection more reliable.

4.8.3 BCMAS Rating and BCARS Rating

The method of defining the building condition scale and failure defer according to the requirement of the client, objective and scope of the inspection. As mentioned in the statement earlier, according to Straub [3], it is important that the building inspector or assessor within organization confirm their own method the same way although the different results of various condition assessment methods in practice.

Table 4.20 Rating scale, representation and inspection for BCMAS

No	Asset type	Condition scale	Linguistic representation	Techniques
1	Buildings	1–3	Condition rating 1 = good—no defect and performing as intended	Visual inspection and functional testing based on approved inspection checklist Data recorded on paper and updating/keep into systems
			2 = unsatisfactory—performing whilst exhibiting defect or deterioration which requires repairs	
			3 = critical—exhibiting defect or deterioration/not operating as intended which needs urgent attention or repair	

Source BCMAS

Table 4.21 Rating scale, representation and inspection for BCARS

No	Asset type	Condition scale	Linguistic representation	Techniques
1	Buildings	1–5	Physical condition 1 = very good, as new, no defect, performing as intended	Visual Inspection and functional testing based on approved inspection checklist (Protocol 1) Data recorded on paper and updating/keep into systems
			2 = good, minor defect, good condition, performing as intended	
			3 = fair, major defect, moderate condition, functioning with supervision	
			4 = poor, major or minor defect, critical, not functioning as agreed service level	
			5 = very poor—very critical, not functioning, risky to safety and health	

Source PWD JKR 21602-0004-13 (2013)

Based on Tables 4.18 and 4.19, the total element assessed in BCMAS is 155 elements with criteria of defects are 134. Meanwhile the element assessed in BCARS which consist of systems, subsystems and components are 551 with generic criteria of defects are 128. Hence, it shows that BCARS is more comprehensive and detailed to make the data collection more accurate and reliable.

Tables 4.20 and 4.21 shows the condition scale used for BCMAS and BCARS. The condition scale used for BCMAS is only 3, which is not representing the condition as overall. This is due to the scope of BCMAS to cater to the cost of rectifying only compared to BCARS. As explained in Chap. 1, the expectation from the result of BCA are for ease of data access, long- and short-term planning for rectifying work and maintenance program, comprehensive budget preparation for maintenance and also for decision making either the building is for upgrading, renovation, refurbish, maintain or demolish.

For BCMAS, the content analysis has been based on the observations and feedbacks from the survey and also from reports. It is recognized that the survey and inspection work carried out are merely visual with no testing on the various building element or components similar as the implementation of BCARS. From the reports, the buildings were categorized by states with regard to the type, category, frequency of the defects, types of elements and discovery. Others is the highest numbers of defects and estimates repair costs were also identified. Defects occurring at the highest frequency are also distinguished. Besides that any significant correlation between the defects, their criticality are also highlighted.

In addition, there is also a similarity in techniques of inspection between BCMAS and BCARS. According to the comparison between BCMAS and BCARS, it seems that BCARS is a more comprehensive assessment tool to the condition of the building toward maintenance. Thus in this study, BCARS will be used to ensure the workability to come out with the prediction model to the building condition. The mapping data of findings then will be validated and compared for the final analysis.

References

1. Rugless, J. (1993). Condition assessment surveys. *Facilities Engineering Journal*, 11–13.
2. Ahluwalia, S. S. (2008). *A framework for efficient condition assessment of the building infrastructure*. University of Waterloo.
3. Straub, A. (2009). Dutch standard for condition assessment of buildings. *Journal of Structural Survey*, 233–235.
4. DFES. (1993). Assessment management plans-condition assessments, section3a: Getting into the condition. Retrieved from http://www.dfes.gov.uk/schoolbuildings.
5. Kaiser, H. H. (1993). *A foundation to uphold: a study of facilities conditions at U.S college and universities, the association of physical plant administrators*. Alexandria, Virginia.
6. INGENIUM, & IPWEA. (2006). *International infrastructure management manual. Thames, New Zealand.*, Association of Local Government Engineering NZ Inc (INGENIUM).
7. INGENIUM. (2006). *International infrastructure management manual*. Public Work Engineering of Australia.
8. CIC. (1996). *Definitions of inspections and surveys of buildings*. CIC.

9. Pedro, C. B. O., & J. A. (2008). Portuguese method for building condition assessment. *Journal of Structural Survey, 25*(34), 322–335.
10. Ramly, A. (2006). *Panduan Kerja-kerja Pemeriksaan Kecacatan Bangunan.* Hizi Print Sdn Bhd.
11. Wordsworth, P. L. (2001). *Building maintenance management* (4th ed). Oxford Blackwell Science Ltd.
12. Pitt, T. J. (1997). Data requirements for the prioritization of predictive building maintenance. *Facilities, 15*(3/4), 97–104.
13. Abbott, G. R., Mc Duling, J. J., Dr, P. S, & Schoeman, J. C. (2007). Building condition assessment: A performance evaluation tool towards sustainable asset management. *CIB World Building Congress*, 649–662.
14. Ismail, Z.-A., Mutalib, A. A., & Hamzah, N. (2015). A case study of maintenance management systems in Malaysian complex and high-rise industrialized building system buildings. *International Journal of Economics and Financial Issues*, 28–35.
15. Building code Queensland 2012, Queensland Government.
16. PWD. (2009). *Dasar Pengurusan Aset Kerajaan (PAK).* Malaysia Public Work Department.
17. Mihai, F., Binning, N., & Dowling, L. (2002). *A Framework for assessing asset management performance in Australia.* Main Roads Western Australia.
18. PWD. (2009). *Manual Pengurusan Aset Menyeluruh Kerajaan (MPAM).* Malaysia Public Work Department.
19. PWD. (2013). *Building condition inspection for existing building.* Malaysia Public Work Department.
20. Kempton, J., Alani, A., & Chapman, K. (2001). Surveyor variability in educational stock survey-a lens model study. *Journal of Facilities, 20*(5/6), 190–197.
21. PWD. (2012). *Tatacara Pengurusan Aset Tak Alih (TPATA).* Malaysia Public Work Department.
22. Kleiner, Y. (2001). Scheduling inspection and renewal of large infrastructure asset. *Journal of Infrastructure Systems, 7*(4), 136–143.
23. Mohseni, H. (2012). *Deterioration Prediction of Community Building in Australia.* PhD unpublished Thesis, RMIT University.
24. PWD. (2013). *Sistem Kod Aset Tak Alih (SKATA).* Malaysia Public Work Department.
25. Georgiou, J., Smith, J., & Love, P. E. (2004). A benchmark study on housing defects in Victoria, Australia. *International Journal of Construction Management*, 89–100.https://doi.org/10.1080/15623599.2004.10773063
26. Hamzah, N., Mahli, M., Che Ani, A. I., Tahir, M., Abdullah, N., & Tawil, N. (2010). The development of smart school condition assessment based on condition survey protocol (CSP) 1 matrix: A literature review. *Science, Engineering and Technology*, 620–625.

Chapter 5
Research Methodology

Abstract Choosing the right research methodology is fundamental to obtain accurate results of the study. The research design carries an important influence in selecting the appropriate research method in the study. An understanding of the research process from initiation of a research idea, thorough literature search, and formulation of a research question, proper study design, and conduction of research allows the researcher to adopt the research design of the study. The present study used triangulation techniques methodology which combined quantitative and qualitative approaches. The triangulation design was used to validate, confirm or corroborate the results findings. The data collected are divided into three stages, including literature review, field building survey for school buildings, analysis for archived documentation and followed by semi-structured interview of 12 professional building surveyors and engineers. Findings results both quantitative and qualitative are further analysed using statistic software Statistical Package for the Social Science (SPSS). The findings and analysis were described in the next chapter.

5.1 Introduction

This chapter described the formulation of the research design and discuss the methodology of the study. The procedures and methods used to conduct the data collection and analysis of data are explained in this chapter. This chapter is presented under the following sections justification for the methodology, research process which consist of determine the research question, proposal of research method, data collection, data analysis, data interpretation, legitimation and the conclusion of the chapter.

5.2 Methodology of Research

The definition of mixed method research by Creswell and Clark [1] states "as a method, it focuses on collecting, analysing, and mixing both quantitative and qualitative data in a single study or series of studies. Its central premise is that the use

S. Yacob et al., *Managing Building Deterioration*, Management in the Built Environment,
https://doi.org/10.1007/978-981-16-5860-0_5

of quantitative and qualitative approaches, in combination, provides a better understanding of the research problems than either approach alone". A mixed method design is useful to capture the best of quantitative and qualitative approaches [1]. Mixed method design also helps to provide more data and more accurate evaluation of analysis. By considering the objectives of the study and the research questions related to building conditions in Malaysia, the study was conducted using the mixed method approach. The reason for adopting this method is because the initial results can be further supported. Other methods used for this study are the use of literature reviews, interviews, and analysis of secondary data using BCA.

5.3 The Research Process

The process of undertaking research is actually the same for pure science, or social science. Research is a systematic search process seeks for the information [2]. Yong and Sulieman [3], described a research involve the procedures such as development of questions, procedures, planning, observation and interpreting to predict other events. Meanwhile [4] quoted from other studies described the different models of the research process and each comes up with different number of stages, but each explained similar steps. In this research, the process of research stages of mixed methods by Johnson and Onwuegbuzie [5] as shown in Fig. 5.1 was followed.

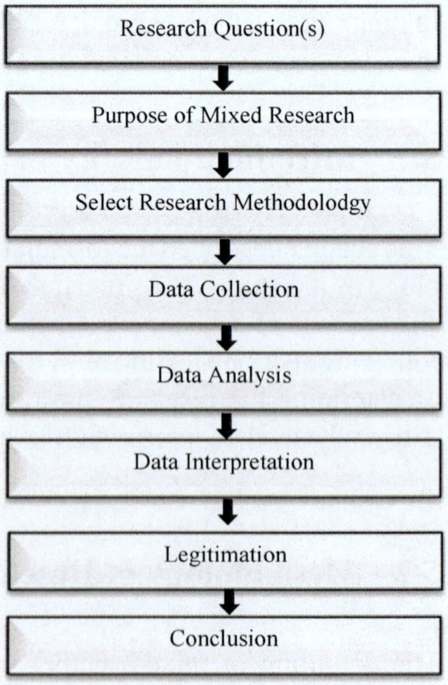

Fig. 5.1 Mixed research process model. *Source* Johnson and Onwuegbuzie (2004)

5.4 Determining the Research Questions

5.4.1 Initial Idea

To seek for the information related to the study, a continuously literature review was carried out which consists of critical reading, evaluating, and organising literature on the topic to assess the state of knowledge in the area. Essentially, these activities involved in the first step [6] that bring the researcher to look for the initial idea to formulate a topic, objectives, questions and research problems. According to Watt [7], a literature search is normally carried out before the start of the study. Literature easily helped to gain the understanding related to the research study Yong and Sulieman [3].

The selection of the research question is a crucial stage, as an inappropriate topic or question may affect the choice of study design, potentially leading to irretrievable difficulties in the research [8]. According to Yong and Sulieman [3], the questions asked in the research provide the important idea and clue to the relevant research approach to be used.

In this stage, the researchers are to decide on the general idea to identify a subject interest that is related to their study. The idea might arise from practice, from professional discussion among colleagues, from reading an article or book, or from an issue in the media. During this stage, the researcher must ascertain the extent of existing knowledge, define their own area of interest to generate with an idea that would be contributing to the knowledge [3].

5.4.2 Identifying the Problem Statement

Researchers must understand the background of the problem encountered with the situation. There are few tasks involved in establishing the problem area. The first one is analysis of secondary data. The analysis of secondary data is important to identify the problem in the study [6]. Using the previous data provided the important clue or information regarding the problem. The second task involved is discussion with related parties involved with the research. By doing this, the actual problems faced are identified and might help the researcher to identify the problem. An exploratory study regarding the background of the problem researched occurred in another country or organization should be involved besides reviewing the literature, which is providing the knowledge, theories, and useful information related to the research. In this study, a review of the literature and exploration of the primary literature and secondary literature from existing knowledge through original results of the articles in journals, dissertation, conference proceeding, review articles, systematic reviews, books, practice guidelines and previous inspection report from Public Work Departments (PWD) to gain better understanding of the study.

The study is then summarized by using cause mapping to get a general idea regarding the current issue to turn into research questions, research objectives, research hypotheses, research topics and research problems to the study. As a result, the research topic is established as "Developing a Building Deterioration Prediction Model (BDPM) for Public Schools Building in Peninsular Malaysia". As explained in the previous chapter, the specific questions addressed in this study are:

(i) What are the factors which affect school building deterioration in Malaysia?
(ii) What is the current state or level of conditions of school buildings in Malaysia?
(iii) To what extent is the relationship between the factors and school building condition?
(iv) What are the factor methods used for prediction based on building condition?

In this research, the author conducted an exploratory study to gather the background of the problem regarding to the building condition before the actual problem is identified in this study. To measure the building condition towards a certain building defects, many possible factors contributed to the building defects that influence the condition of building. There are a lot of discussions about the contributory factors to building defects or building failures especially during construction and pre-occupancy stage. There are few discussions between factors contributing to building defects and building condition by using building condition assessment tools for existing building.

Subsequently, the research objectives were formulated for this study as follows:

(a) To identify the factors contributing towards building defects;
(b) To investigate the condition of public-school buildings through building condition assessment;
(c) To establish critical factors contributing towards school buildings defects and its' relationship to building condition;
(d) To develop a building deterioration prediction model (BDPM) of the building condition.

To isolate and identify the most likely causes based on the background of the study, was developed based on literature and secondary data as shown in Fig. 5.2.

5.5 Proposed of Mixed Method

Mixed method is defined as concurrent or sequential use of a single method that include quantitative and qualitative component [9]. The reasons why this method was preferred because mixed method efficiently generate greater understanding. In this study, mixed method was proposed because of the results from quantitative measures need to be confirmed with qualitative experiences. Johnson and Onwuegbuzie [5], reported that mixed method techniques within the same framework can incorporate the strength of both methodologies. For this research, the mixed model adopted in

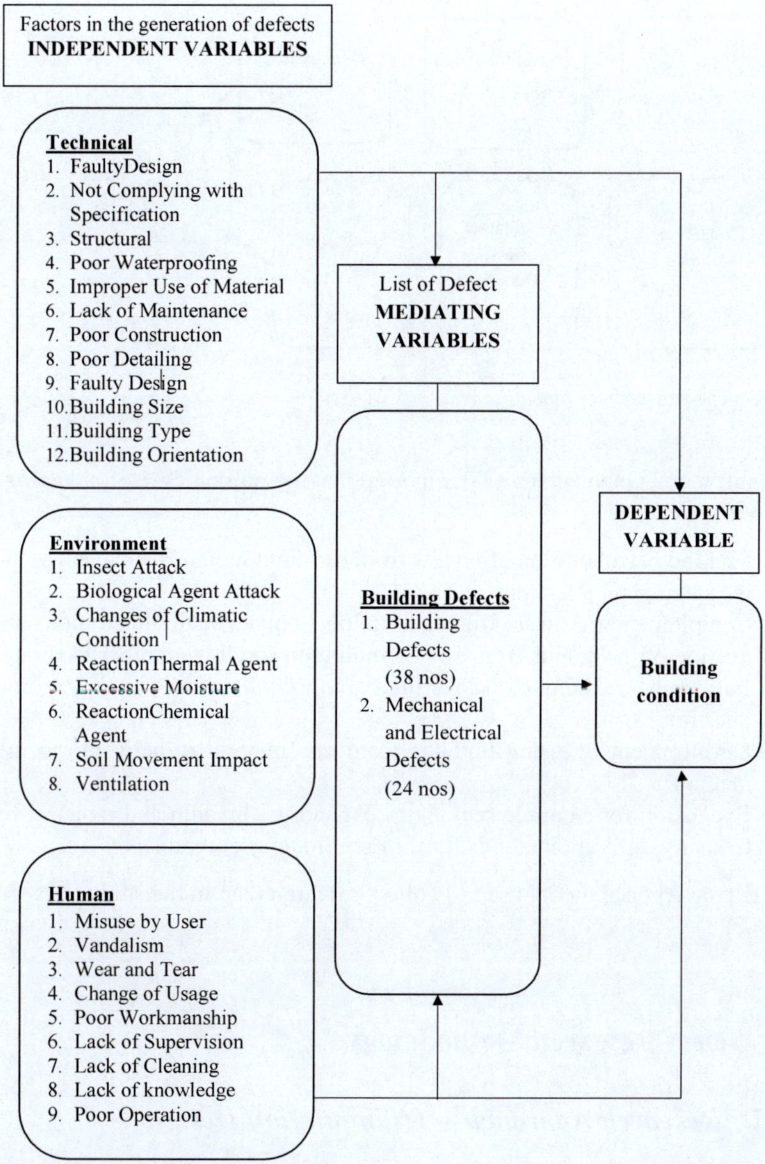

Fig. 5.2 Theoretical framework

the study is convergence model. Creswell and Clark [1] mentioned that the model is actually a mixed methods triangulation model.

According to a study by Mwaniki and Mue [10], in this model the researcher collects and analyses quantitative and qualitative data separately on the same phenomenon and the result are converged during interpretation. Similar to studies

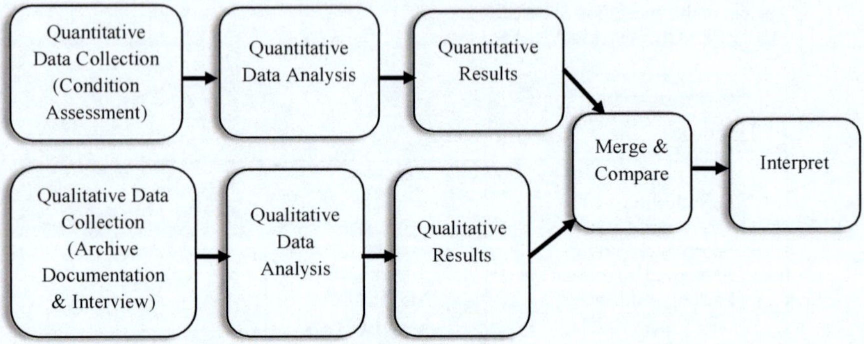

Fig. 5.3 The convergence model

by Johnson and Onwuegbuzie [5], reported that the rationales for conducting mixed methods which are:

(a) Seeking corroboration of results from different methods and designs studying the same phenomenon;
(b) Complementarity in seeking elaboration, enhancement, illustration, and clarification of the results from one method with result from other method;
(c) Initiation for example discovering paradoxes that lead to a re-framing of the question;
(d) Development by using findings from one method to help inform the other method;
(e) Expansion for example seeking to expand the breadth and range of research by using different methods for different inquiry components.

The convergence model used to collect research data in the study is as shown in Fig. 5.3.

5.6 Select Research Methodology

5.6.1 Research Approach—Triangulation Design

A triangulation design was used in this study. This triangulation design was used to confirm and validate or corroborate quantitative results with qualitative findings. Mwaniki and Mue [10], explained that triangulation is the act of combining several research methods to study one thing and it is used to indicate two (or more) methods in a study in order to check the result. Yeasmin and Rahman [11], explained that the use of triangulation is to enhance understanding under investigation by combining multiple methods and theories. By combining methods in the same study, observers can partially overcome the deficiencies that flow from one investigator or method.

They also explained that the advantage of designing multi-method research lies in the potential for enhancement of the validity of the study findings.

This study is designed based on the research design adopted by Ali [12] in his study to introduce an integrative mechanism to improve the performance through the relationship between the uncertainties of refurbishment design process. The methodology of this research also referred to the study by Yong and Sulieman [3] to study the relationship between the preventive maintenance characteristic of the maintenance performance of the high-rise office building in Malaysia to develop a prediction model for maintenance performance. Both of them used the triangulation design in their studies. This study also referred to the study by Ahluwalia [13], regarding to efficient condition assessment of the building by addressing problems associated with the condition of the buildings.

In this study a condition assessment was conducted to collect the quantitative data. And then for the collection of qualitative data, archive documentation and the interview were conducted and analyse to validate the quantitative results. According to Ali [12], validity in research is about the strength of research conclusions. He also explained that if the investigation is conducted correctly, a better understanding of the relationship between hypotheses can be obtained. Table 5.1 shows the research design formulated in this study.

Research by Fox [14] stated that triangulation is important for the verification of results since the data collected from a single source could introduce the element of bias due to self-justification. For this research, a diagram of author design adopted the convergent design by Wittink et al. [15] as shown in Fig. 5.4.

5.6.2 Population and Sampling

5.6.2.1 Population

A population can be defined as the complete set of subjects that can be studied. The population as the entire group or set of cases that researcher is interested in generalizing [16]. For both quantitative and qualitative methods, the population selected in this study were listed in the Directory of Federal Government Premises Registration, [17], published by Public Work Department (PWD) Malaysia.

The study population of the overall number of Government buildings registered in Malaysia are 13,356 as explained in Chap. 2. Private or independent building is not included in the population. The total number of education building is 10,381 consisting of 161 universities and colleges, 113 for kindergartens, 37 training centres, while 10,070 are school buildings.

Table 5.1 Research design

Research objectives (RO)	Data collection method	Data analysis method	Respondent (if applicable)
RO 1 To identify the factors contributing towards building defects	(a) Literature review – Reviewing previous research – Documents, and journals, articles related to the study (b) Archived documentation – BCMAS reports to identify factors contributing to building defects – Define the sample type, numbers, age, location of the school	(a) Descriptive analysis (b) Content analysis	Expert from PWD to validate the group of similarity of factors identified used in the study – Independent variables
RO 2 To investigate the condition of public-school buildings through building condition assessment	(a) Quantitative method – BCA uses building condition inspection of existing building guideline (JKR 21602-0004-13) (b) Qualitative method—archived documentation – BCMAS report to identify the extent/degree of defects (c) Qualitative method – Semi-structured interviews to verify, confirm and validate the result of findings	(a) Descriptive analysis (b) Reliability analysis (c) Content analysis	Expert from PWD to validate the theoretical framework and results of findings – Mediating variables – Dependent variables Also to confirm the practical use of BCA Guideline
RO 3 To establish critical factors contributing towards school buildings defects and its' relationship to building condition	(a) Results of RO1 and RO2 – Independent variables – Dependent variables (b) Qualitative method – Semi-structured interviews to verify, confirm and validate the result of the findings	(a) Correlation (b) Content analysis	Expert from PWD to validate the results of findings

(continued)

Table 5.1 (continued)

Research objectives (RO)	Data collection method	Data analysis method	Respondent (if applicable)
RO 4 To develop a building deterioration prediction model (BDPM) of the building condition	(a) The result findings of RO2 and RO3 – Merged result to come out with BDPM	(a) Logistic Regression	None

5.6.2.2 Sampling Frame

For this study, the school buildings were selected as a sampling frame according to a set of criteria based on the characteristic of the population. The sample population for the school buildings drawn from the sampling frame is 10,070. The population known as sampling frame and can be thought of as a list of all people relevant to the study [14]. Cobbinah [18] explains that a sample frame is a list that includes every member of the population from which sample is to be taken. The reason for selecting school buildings as according to Andriotis [4], the school and educational buildings was the largest infrastructure sectors and most of them were built more than 45 years old and need for extra care.

Previous research has shown that the highest number of building owned by government in Malaysia was school building and as audited and reported are in poor condition, hence the school maintenance is still practiced in an unsatisfactory level, which lead to the increment of the maintenance cost of school buildings [3, 12, 19]. According to [20], there are numerous defects that arise and being reported officially by mass media is educational buildings. Mahli et al. [21], noted that the school is one of the government's biggest asset and should be maintained to ensure its functionality and value. Meanwhile [21], stated the condition of school building in Malaysia were not in good condition. For the accessibility of the population chosen is in the locale in Malaysia and based on previous study by NCES [22], and [23]. From the school sampling frame, a few criteria set up based on previous studies related to the school buildings are presented in Table 5.2.

5.6.2.3 Sampling

According to [18], explains that the sample is a sub-group of the population which is representative of the population. Appropriate sample size is required to represent the research population [24]. The quality of the sample affects the research generalization. The size of sample will be determined based on the scope of the study. The type of sampling used in the study is stratified random sampling. From the sample frame of school buildings in Malaysia, the total number of schools will be minimized to a minimum sample size based on school age, level and region.

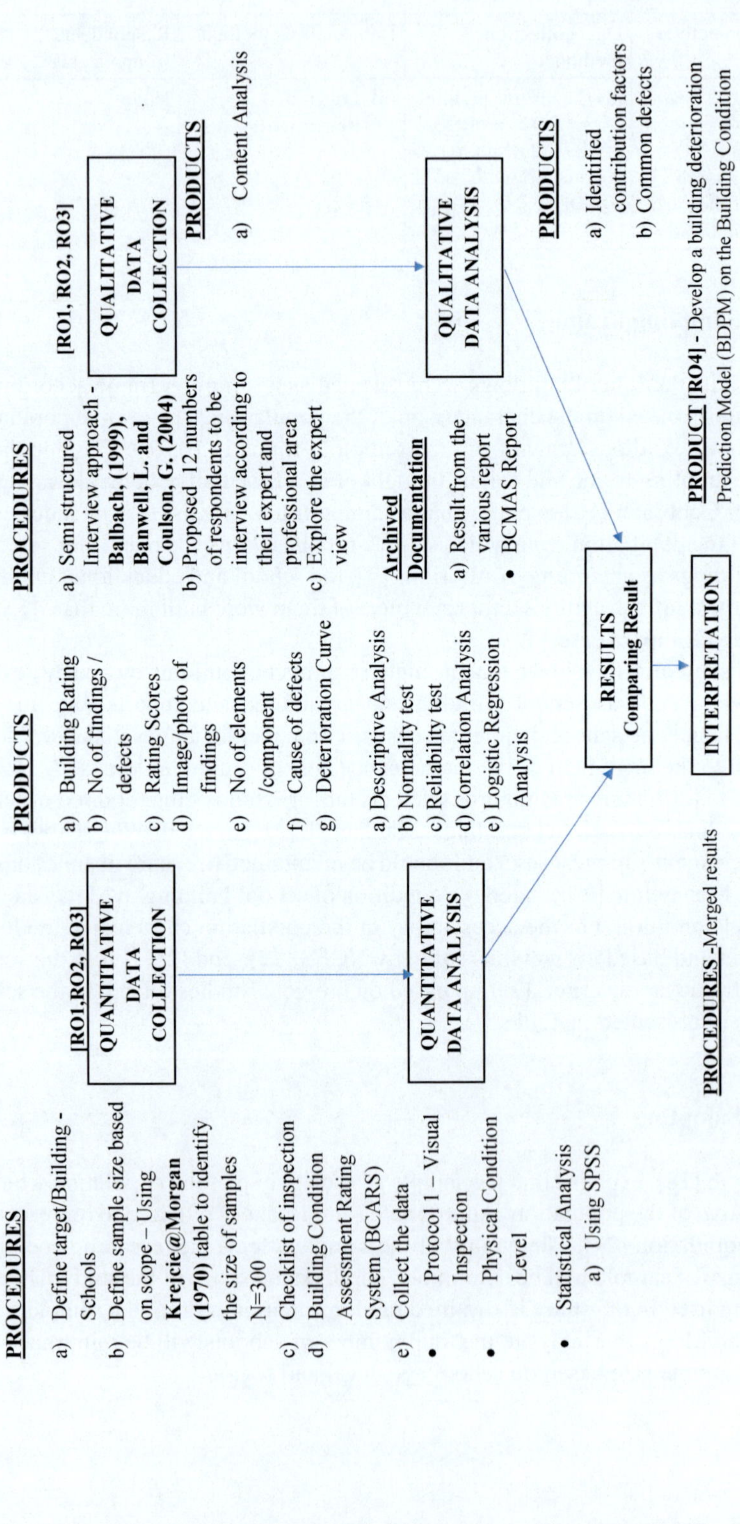

Fig. 5.4 Covergent design for the study

Table 5.2 Criteria for target sampling frame for school building

Population	
Target population: school buildings in Malaysia (10,381)	
Criteria/	Literature review
• Highest total number of buildings owned by the government • Schools in Malaysia audited were in poor condition • The increment of maintenance budget for school buildings • School maintenance is still practiced in an unsatisfactory level • BCMAS records of Federal Government Buildings that were kept by PWD	– MySPATA report for 2011 (JPAK) vAli, Keong, Zakaria, Zolkafli, Akashah (2013) – Yong, Zailan, Sulieman (2015) – Report of Overall Trend Assessment & Analysis Report (2011) –Kumpulan IKRAM Sdn. Bhd
Accessible population: school buildings based on locale & region	
Criteria	Literature review
• Locale & Region	– NCES 2000–032 (National Centre for Education Statistics) – Report of Overall Trend Assessment & Analysis Report (2011)—Kumpulan IKRAM Sdn. Bhd – Mahli, Che Ani, Abd-Razak. Tawil, and Yahaya (2012) – RMK 9—2006 to 2010

Through research, it is found the focus requirement for the scope of study is school age range from 10 to 20 years old. According to [21], the school age ranged from 1 to 65 years and critical school buildings age is between 10 and 20 years. And the school selected mostly built of reinforced concrete frames. According to [25], found that most of the building defects become visible 2 years after occupancy. The building needs to be repainted once every 5–6 years. They also reported that the defect data were collected from buildings between 2 and 6 years old. Meanwhile according to [22], the functional age of the school is 16 years.

School institutional levels consist of primary and secondary school with two types of schools i.e. National School and National-type School. According to [17] other types of schools registered are Cluster School, Vision School, Islamic Religion School, Mubaligh School, Bestari School, Vocational and Technic Schools, and Full Boarding School. According to Ali [12], reported that design defects for National schools and National-type schools have caused high maintenance cost at state of Kedah. Research by [26], also focus on primary school with National School and National-type School at state of Perlis. Lewis and Payant [27] suggests that the larger the sample size the greater the probability the sample will reflect the general population.

As such, the author decided the location of the school buildings be chosen in this study are North area which covered the state of Kedah, Pulau Pinang and Perlis. For East area, the states covered are Kelantan, Terengganu and Pahang. South Area covered the state of Johor and Melaka and for West area are Perak, Selangor and Negeri Sembilan. Previous research has shown that the location, such as central city, urban fringe / large town and rural/small town was covered in the research [21, 23]. The school institutional level covered include both primary and secondary with types of National School and National-type School for 11 states in Malaysia not including Sabah and Sarawak were selected for this study.

Table 5.3 shows the summary of the sample criteria selection of the study. The schools age that was selected was within the age of between 6 and 10 years old, starting when the documentation was prepared. This is important to ensure the research objectives of the study to identify the factors contributing towards building defects for schools can be identified. At that particular age, defect data can be collected from buildings between 2 and 6 years old [25]. The types of schools selected were National and National-type school at Peninsular Malaysia with reinforced concrete structure. The conventional school buildings designed by PWD are mostly built of reinforced concrete frames. The investigation on the conditions of the school buildings using BCA is necessary to achieve the second research objective of the study. The analysis of both research objectives 1 and 2 findings will generate data for the research objectives 3 and 4.

5.6.2.4 Sampling Size Determination

Based on filtration process the total number of schools found located in the states of Malaysia (see Appendix B) fulfilled the scope of study which is $N = 974$. According to [24], the population $N = 1000$ is used in this study, which is close to the scope of the study. As shown in the table the recommended sample size representative of the school building in this study should not be less than 278. However, Baker and Edward [28] stated that for statistical validity, the larger the sample, the more accurate will be the results. The author wants to get as many cases as having different conditions related to the phenomenon under study. The advantages of a large sample size to interpret significant results are it allows a more precise estimate and also to generalize the results.

Therefore, the sample size used in this research is 303. The total of 303 schools is as per Directory of Federal Government Premises Registration. The registration of schools was according to the premises which included for several blocks of building.

5.7 Data Collection

As explained earlier, this study adopts the quantitative and qualitative method. Thus, the data from each method gathered are merged and compared to come out with the interpretation of the results. Data collection is important in a research study. Watt [7], defined data as information obtained during investigation or study.

Table 5.3 The criteria selection for scope of study

No	Criteria	School characteristic	Focus requirement	Literature	Research objectives (RO)
1	School age (years)	a. School age is range 1 year to 65 years b. The older the school, the more building defects occurred c. The critical school buildings age is between 11 and 20 years	• 1–5 • 6–10 • 11–15 • 16–20 • >20	• Mahli, Che Ani, Abd-Razak. Tawil, and Yahaya (2012)	**RO 1** To identify the factors contributing to the building defects **RO 2** To investigate the condition of the public- school buildings through building condition assessment **RO 3** To establish critical factors contributing to the school buildings defects and relationship to building condition
		a. Building defects visible 2 years after occupancy and building to be repainted once every 5–6 years	Defects were usually rectified during re-painting; defect data were collected from buildings between 2 and 6 years old	• Wai-Kiong Chong, M.ASCE, and Sui-Pheng Low (2006) – Records, Interviews and guided walks	
		a. Average age of public school is 40 years b. Years since most recent renovation are 11 years old c. Functional age of the school is 16 years	• Less than 5 years old • 5–14 years old • 15–34 years old • 35 or more years old	• NCES 2000–032 (National Centre for Education Statistics)	

(continued)

Table 5.3 (continued)

No	Criteria	School characteristic	Focus requirement	Literature	Research objectives (RO)
2	School level	a. Primary b. Secondary	**School enrollment size** • Less than 300 • 300–499 • 500–999 • 1500 and more **Types of school** • National school • National-type school **Types of constructions** • Reinforced concrete Conventional buildings are mostly built of reinforced concrete frames	• Ministry of education • NCES 2000–032 (National Centre for Education Statistics) • li, Keong, Zakaria, Zolkafli, Akashah (2013), Yunus, (2013) • MySPATA—*Sistem Pengurusan Aset Tak Alih* • Andres, C.K, and R.C Smith. 1998. *Principal and Practices of Heavy Construction*. 5th Edition, New York: Prentice Hall • Haron et al. (2005)	**RO 1** To identify the factors contributing to the building defects **RO 2** To investigate the condition of the public- school buildings through building condition assessment **RO 3** To establish critical factors contributing to the school buildings defects and relationship to building condition
3	Locale & Region	Kedah, Pulau Pinang, Perlis, Kelantan, Terengganu, Pahang Johor, Melaka, Perak, Selangor, Negeri Sembilan	• Central City • Urban fringe/large town	• Report of overall trend assessment & analysis report (2011)—Kumpulan IKRAM Sdn. Bhd • Mahli et al. (2012) • RMK 9—2006 to 2010	

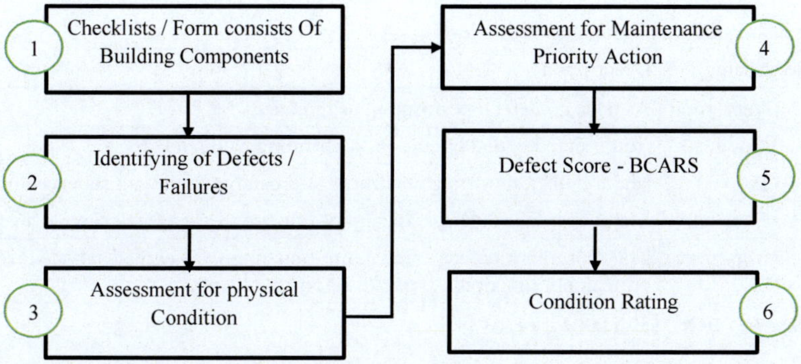

Fig. 5.5 Condition assessment process. *Source* PWD JKR 21602–0004-13 (2013)

5.7.1 *Quantitative Method*

The quantitative approach for this study generated, data collected using BCA to understand of the condition and capacity of the building and to identify the current condition of the building. The approach of the research is quantitative research which is well-suited for the study to testing the outcomes of the BCA. According to Andriotis [4], quantitative data are drawn to quantify data by applying mathematical models, graphs and statistical tables to achieve more reliable results.

5.7.1.1 Condition Assessment

In this study, for assessing the building condition for school buildings, the guidelines and systems the Building Condition Inspection for [29], were used and is explained in detail in Chapt. 3. Figure 5.5 shows the method of data collection according to the BCA guideline used in this study.

5.7.1.2 Checklists/Form Consists of Building Components

In this study, the collected data was referring to the checklist. The data was compiled for registering the types and others information of the defects. The checklist consists of five (5) parts of assessment which is the information of the building, civil and structure, mechanical, electrical and external work checklists.

5.7.1.3 Identifying of Defects/Failures

The findings identified are recorded in the defect's sheets. Then the defects are assessed and illustrated with photographs to give a real condition of the component.

Table 5.4 Building physical condition level

Grade	Scale	Description
1	Very good	As new, no defect, performing as intended
2	Good	Minor defect, good condition, performing as intended
3	Fair	Major defect, moderate condition, still can function with supervision
4	Poor	Major or minor defect, critical, not functioning at agreed service level
5	Very poor	Major or minor defect, critical, not functioning as agreed service level, very critical, not functioning, risk to safety and health

Source PWD JKR 21602-0004-13 (2013)

The detailed inspection of the defects to be filled and checked on readiness condition damage or defects found, and priority defects or damage needs to be given priority. The contributing factors to those defects also were identified according to research framework.

5.7.1.4 Assessment for Physical Condition

BCA rating commonly uses a weighting factor to represent the severity of each defect. The assessment based on scale building physical condition level as explained in Chap. 4. Table 5.4 shows examples of condition scales and its representations as used in the standard guideline of BCA.

5.7.1.5 Assessment for Maintenance Priority Action

Then, the building physical condition rating will link to the maintenance priority actions and types as shown in Table 5.5. The analysis of the defect findings links to the defect parameter as shown in the matrix analysis in Table 5.6.

Table 5.5 Maintenance priority action

Priority	Scale	Description
Normal	1	No defect or damages, element/component well maintained
Routine	2	Minor defects/damages, needs for monitoring, repairs, replaced to prevent serious defect/damages
Repairs	3	Major defects/damages, needs for major repairs and replacement
Rehabilitation	4	Critical/serious defects/damages, needs for urgent and immediate repairs
Replacement	5	Critical/serious defects/damages, needs for urgent replacement, refer to expert detail inspection/Expert judgement

Source PWD JKR 21602-0004-13 (2013)

Table 5.6 Matrix analysis building physical condition and maintenance priority actions

Scale		Maintenance priority action				
		5	**4**	**3**	**2**	**1**
Building physical condition	**5**	25	20	15	10	5
	4	20	16	12	8	4
	3	15	12	9	6	3
	2	10	8	6	4	2
	1	5	4	3	2	1

Source PWD JKR 21602-0004-13, (2013)

5.7.1.6 Defect Score—BCARS

In this research, the defect and the maintenance priority action combined with a defect score, in BCARS. The results were analysed and delivered to summary of form to the conclusion of quantitative results. The overall building rating derived from the total mark from sum up matrix analysis were divided by total number of defects.

5.7.1.7 Condition Rating

Table 5.7 shows the five-point colour-coded building rating is shown as a final result of the assessment. The building condition with a rating very good condition (A), range score is 1–5 with the action preventive maintenance. The good condition (B), range score is 6–10 with condition-based maintenance required. The fair condition (C), range score is 11–15 which require for repair work. The poor condition (D), range

Table 5.7 Overall condition rating

Rating	Physical condition	Action matrix	Score
A	Very good	Preventive maintenance	1–5
B	Good	Condition based maintenance	6–10
C	Fair	Repairs	11–15
D	Poor	Rehabilitation	16–20
E	Very poor	Replacement	21–25

Source JKR 21602-0004-13 (2013)

score is 16–20 required for rehabilitation work. Meanwhile the very poor condition €, range score is 21–25 needs for replacement work.

5.7.2 *Qualitative Method—Archived Document and Semi-structured Interview*

Silva [30], explains that, the strength of qualitative study is to stress on the importance of context, setting and participants' frames of reference. Yong and Sulieman [3], stated that qualitative method is to merge, compare and validate the quantitative results. Yong and Sulieman [3], also highlighted that the quantitative method is the main investigation relied for the study. In this research, qualitative method involved were semi-structure interview and analysis of archived documentation. The results will be compared and merged with quantitative method which is observation through building survey.

5.7.2.1 Archived Documentation

It is important to study and understand the data and/ or the information of archived documentation. In this study, archived documentation was carried out to identify the current building condition and to identify the contribution factor to defects. Yong and Sulieman [3], adopted qualitative method which is archived documentation in his study to examine records and identify the categories of events related to the study. Othman et al. [31], conducted a research based on secondary data by analysing all related documents to find out the building defects. Graziano and Raulin [2], stated that archived document provided related information about the event. Using archived documents or existing data in the evaluation of the study constitute secondary data analysis. For this study based on secondary data was adopted.

In this study, the content of all reports has been analysed to identify factors contributing to the defects and assessing the defects of the building. Yong and Sulieman [3] in his study used existing records to determine variables. The variables (factors) identified from the literature will be used in the study to determine the variables to the study. Megha and Rajiv [32], suggests Relative Importance Index (RII) is a method to determine relative important of the various causes and the effects of the factors studied. The same method was adopted to use in this study. The factors identified have been analysed using relative index (RII) to determine the relative importance of the factors causing the building defects (see Appendix C).

The factors affecting the defects in the buildings were scored on a five-point Likert Scale based on their significant. The scale for rating the factors are five (5) for very high, four (4) for high, three (3) for average, two (2) for low and one (1) for very low. From the archived documents, the degree of defects was based on BCMAS rating scale assessed by professionals who prepared the previous reports. The assessment in this study was done by observing the existing rating scale suited with the degree of defects in the reports. The assessment was performed on photos, and the factors contributing towards the building defects.

$$\text{RII} = \frac{\sum \text{W}}{\text{A} * \text{N}}$$

Where, W is a weighting given to each factor which contribute to building defects (ranging from 1 to 5), A is highest weight (in this study is 5) and N is the total number of defects found in the analysis for each building assessed. The Higher the value of RII the more important [32], was the cause of defects. They also noted that the RII was used to rank the different causes. RII should be ranked in order to give the overall picture of the causes which contribute to the defect of the buildings.

Pershing [33], reported the purpose of the document analysis whereby document is any archived source of information. Firstly, in this study, the purpose of the document analysis is to fulfil the objectives of the study. Secondly, a total number of the 303 school buildings were selected from the archived documentation. The criteria selection for sampling size for schools is the same as the criteria set up in sampling for quantitative data. The archived document selected was from the current building inspection report that was kept by PWD. The reports of PWD is Building Condition Management Assessment System report. All the defects of the school building can be obtained in this report. All school building, then were recorded by name and coded according to their state. In this study also the researcher formulated a questionnaire used in semi structured interview.

In this study however, the data collected from archived documents, were reports that were quite outdated. The archived documents that was assessed is from year 2012 which means, the age of schools assessed is about 6–10 years. The selection of schools are still within the area of scope of study since the building defects was reported visible after 2 years of occupancy [25].

5.7.2.2 Semi-structured Interview

To obtain more detailed information for this study, interviews with professionals will be conducted. To ensure that the data collected is accurate and reliable, respondents will be selected according to their expertise and experience area. In this research, the semi-structured interview approach was selected as it gives the interviewer more flexibility to explore views, ideas and opinions while maintaining a level of comparability between interviews. This method is needed to enhance a primary method and making the findings more concrete.

Interview is dominant in the field of qualitative research [34]. According to Wilson [35], interviewing is used for gathering information about task, discussion reaction to the research findings and etc. Yong and Sulieman [3], suggested the interview conducted was to gather more information regarding to the respondents' views on the phenomenon of interest instead of depending on the literature and archived documentation. In this study, semi structured interviews will be carried out to obtain more detailed information and to validate the results of quantitative data. Cohen and Crabtree [36], explains that semi-structured interview provides a clear set instruction for interviewers and can be reliable, comparable qualitative data.

Yong and Sulieman [3], quoted that the used of semi structured interview to seek more understanding about the relationship between variables. Meanwhile, according to Wilson [35], a semi structured interview to be conducted when issues under investigation need further details for clarification or validation. The relevant respondents must be selected according to their professionalism and expert in the field in order to obtain accurate and reliable data. In this research, a semi-structured interview approach was selected as it allowed the interviewer more freedom to explore the expert views or opinions while maintaining a level of comparability between interviews.

According to Baker and Edward [28], there are no set guidelines on how many interviewers are enough and the sample size for the interviewer depends on the methodology, research questions and context. Stuart et al. [37], advice the number of interviewers can vary from 5 to 35 interviewers. In this study, data gained by analysing the secondary data which is the current building inspection report. The document analysis from the archived report is concerned to identify the current condition of the building and various factor which contribute to building defects. The questions based on the findings were developed to validate the results as follows:

(i) Factors which have strong significant and affect the overall condition of the building;
(ii) Factors which have least significant and affect the overall condition of the building;
(iii) Factors which have zero variance to the overall condition of the building;
(iv) Validation of theoretical framework;
(v) The uses of Standard Building Condition Assessment in the study.

In this study also, saturation was reached after twelve (12) number of participants were interviewed with similar answer without new information. In this study, the participant's criteria set up to ensure the best response are:

(a) A participant for this discussion must be a professional level and as building inspector for BCA in PWD which have working experience of more than 3 years in conducting BCA for government buildings.
(b) The background of the participant may vary, and targeted participant will be Building Surveyor and Engineer at PWD
(c) The participant will be informed for the session of discussions and bound with consent form for participation.

The information was prepared before conducting the interview session. According to Saunders et al. [38], the relevant information must be supplied before the interview session. In this study, the information prepared related to the research for validating the finding of quantitative data is represented in Appendix D. Yong and Sulieman

Table 5.8 Coding of the interviewer

Discipline background	Coding of participant	Requirements
(1) Building surveyor	Participant BS1 to BS6	• Professional—technical or management • Building inspector • Working experience >3 years in BCA
(2) Engineers	Participant E1 to E6	

[3] highlighted the information should be included for the interview and all feedback are confidential and be used for study purposes only. In this study, the information consisted of a brief explanation of the study, aims, requirements and confidentially. The potential numbers participants selected for this study, according to requirement mentioned above is twelve (12) participants. PWD professionals such as Building Surveyors and Engineers selected are selected as interviewees for this study. They were selected because they were appointed as inspectors to enforce the BCA guideline [29] with the jurisdiction of the government of Malaysia. Table 5.8 shows the coding of interviewee based on their discipline.

The interviews end for minimum 30 min, according to the questions set up. According to Georgiou et al. [39], the interview is conducted individually or in group for a duration of 30 min or more. Each interview was recorded and then transcribed into Word. MacLean et al. [40], recommended the researcher replay the record to clarify audible responses.

5.8 Analysis of Data

After all the data are completely collected, such data would be analysed and studied to produce an outcome for the research. The data obtained were analysed using Statistical Package for the Social Science (SPSS) software. Selected appropriate statistical tests were employed using the software, based on the types of data and approach of hypothesis testing [41].

5.8.1 Quantitative Analysis

5.8.1.1 Descriptive Statistics

Descriptive statistic describing the data findings and it is used to present a quantitative description in manageable form [42] and provide important information about variables to be analysed [43]. In other words, descriptive statistic helps to reduce the data into simpler summary and easy to interpret. Hence, Gravetter and Wallnau [44], concluded the consequences of doing the experiment is to end up with a score or set of scores for each individual variable. Trochim et al. [42], suggests analysis involved the examination across cases of one variable at time is univariate analysis which is frequency distribution, central tendency and dispersion. Graziano and Raulin [2] highlighted the descriptive statistics measure of central tendency, variability, and relationship of the data. Thus, according to the objectives of the research, these techniques are suitable and was adopted to analyse the research data for the following reasons:

(i) **Frequency Distribution**

Quantitative data collected for this study are nominal and ordinal data which involved condition rating data and RII and rank, so frequency is able to describe a single variable of the distribution of school building assessed according to their conditions. Frequency distribution is depicted in graphs or tables.

(ii) **Central Tendency**

Central tendency is to measure mean, median and mode of the variables. According to Trochim et al. [42], central tendency of the distribution is an estimate of the centre of the distribution of values. Meanwhile, Park et al. [43], noted that mean, median and mode measure the central tendency of a variable. In this study, this analysis was adopted to analyse the data collected to fulfil the objectives of the study. The objective of the study is to establish the relationship between contribution factors towards building condition and to develop building deterioration prediction model (BDPM) for building. Mean is to measure average score in a distribution. Gravetter et al. [44], explained that mean is computed by adding the set of score and divided by the scores as formulated as:

$$\overline{X} = \frac{\sum X}{N}$$

Where ΣX = sum of the X scores, N = total number of scores.

(iii) **Dispersion**

Dispersion refers to the spread of the values around the central tendency [42]. Standard deviation is common to measure dispersion. Foster et al. [45] standard deviation is used when mean is being used in the analysis. Trochim et al. [42], explains that the standard deviation shows the relation that set of score has to the mean of the sample. The bigger the standard deviation, the more the score differ from the mean. That means the more spread of data shows the randomness of the data and eliminates the bias issue [3]. The standard deviation formula used as estimate to the population standard deviation is as below:

$$s = \sqrt{\frac{\sum (X - \overline{X})^2}{n - 1}}$$

Where, x = each score, x-bar = mean; n = total number of values; Σ means we sum across the values.

5.8.2 Normality Test

Normality test is necessary to be performed before applying a method that assumes normality and objectively assess the data comes from normal distribution or skewed to one-sided [41]. Ghasemi and Zahediasi [46], noted that normality to be considered for validation of the data presented in the literature whether correct statistical tests have been used. Ali [12], noted that this test is applicable where the decision to use either parametric or non-parametric methods. In this study, it is assumed that the data under test is normal distribution [46, 47], explained that the statistical procedures such as regression, t tests, and analysis of variance are based on the assumption that data follows a normal distribution.

5.8.3 Reliability Analysis

Reliability refers to the consistency and stability in the results of a test [48]. It is important [3], to use a reliable scale in the study [12]. To test the reliability of the data, Cronbach's alpha is used in this study. According to Tavakol and Dennick [49], alpha is an important in evaluation of assessment and mandatory to the researcher to add validity and accuracy to interpret their data. It is the same analysis adopted by [3] in his study to check the reliability of his data. Cronbach's alpha is a measure of internal consistency [50].

Tavakol and Dennick [49], explained that many reports about the acceptable values of alpha. The acceptable value of alpha ranging from 0.70 to 0.95. This support statement by [3] in his study noted that the value of alpha that reflects acceptable reliability should be above 0.70. Meanwhile [51] indicated alpha value ranging is 0.60–0.80 and 0.60 is considered poor reliability. But [52], argued that alpha value range 0.60–0.69 is acceptable when there are only items in scale. In this study, the reliability analysis was conducted to test reliability of the variables collected (see Appendix E), Cronbach's Alpha coefficient for characteristic of factors which contribute to building defects is 0.713 and indicates that the scale and data obtained are reliable.

5.8.4 Bi-Variate Analysis

Ali [12], stated that Bi-variate analysis refers to the correlation test of two variables. Correlation is a statistical technique that is useful to determine if there exists a linear relationship between two variables [45]. It is also a useful tool to show how strong or significant is the relationship between the variables. The correlation coefficient or "r" is normally used to measure the strength of the relationship and reported at a value between -1 and $+1$. A correlation coefficient, r value close to -1 or $+1$ showed a strong relationship between the two variables and the r value close to 0 indicates little or no relationship between the two variables [53].

The correlation tests used in this study is the Spearman Rank-Order Correlation Coefficient Technique. McDonald [54], noted that Spearman correlation is to test the association between two ranked variables, or one ranked variable and one measurement variable. The Spearman rank correlation was used for two ranked variables, to see whether the two variables covary; whether, as one variable increase, the other variable tends to increase or decrease. Ali [12], stated that it is non-parametric method of correlation coefficient and normally measures data on at least an ordinal scale. In this study, Spearman rank correlation was used when condition rating as one measurement variable and one ranked variable such as factors which contribute to building defects.

5.8.5 Spearman Rank-Order Correlation Coefficient Technique

McDonald [54], noted that Spearman correlation is to test the association between two ranked variables, or one ranked variable and one measurement variable. According to [12], it is non-parametric method of correlation coefficient and normally measures data on at least an ordinal scale.

5.8.6 Regression Analysis

Regression analysis is a statistical tool for the investigation of the relationship between variables [14, 55]. The objective of regression analysis is to describe the relationship of a response with explanatory variables [56]. Regression analysis was conducted in this study after correlation coefficient had quantified. Regression analysis was used to indicate the significant relationship between dependent and independent variables and to indicate the strength of the impact of multiple independent variables on a dependent variable. Yong and Sulieman [3] quoted the study from Graziano and Raulin [2] which suggest a strong relationship between two variables provides information by knowing the value of one variable to predict on the other variable.

The regression equation is written as follows:

$$Y = a + bX$$

where,

(a) Y = dependent variable (target/response) what is being predicted or explained;
(b) a = constant; equals the value of Y when the value of X = 0;
(c) b = the coefficient of X; change in Y for a change of one unit in X;
(d) X = independent variable (predictor), what is predicting or explaining the value of Y.

Multiple regression is an extension of linear regression. It is used to predict the value of a variable based on the value of two or more other variables. McDonald [54], stated to use multiple regression when the data have more than two measurement variables. The purpose of multiple regression is to find an equation that best predict the dependent variable as linear function of independent variables [54]. Yong and Sulieman [3], explained that affect the dependent variable would be able to make a more accurate prediction. The multiple regression equation is written as follows:

$$Y = a + b_1X_1 + b_2X_2 + \ldots + b_nX_n + \varepsilon$$

where,

(a) Y = dependent variable (target/response) what is being predicted or explained,
(b) a = constant; equals the value of Y when the value of X = 0;
(c) $b_1, b_2, \ldots, b_n$ = the coefficient of X; change in Y for a change of one unit in $X_1, X_2, \ldots, X_n$ respectively.
(d) $X_1, X_2, \ldots, X_n$ = independent variable (predictor), what is predicting or explaining the value of Y.
(e) ε = random error.

In order to select an appropriate type of regression analysis, the researcher must have assumptions of the regression method selected to check whether the data that the researcher want to analyse can actually be analysed using regression method selected. Most of statistical tests rely upon certain assumptions about variables used in the analysis [57].

In this study, logistic regression analysis was used in the study and the data used was non-parametric data. Logistic regression is used to predict the outcome of being a case based on the values of the independent variables. Logistic regression is used to describe data and to explain the relationship between one dependent binary variable and one or more nominal, ordinal, interval or ratio-level independent variables. Yong and Sulieman [3], noted that logistic regression is similar to multiple regression and it was used in his study to predict the maintenance performance involved a set of independent variables. Sekaran and Bougie [51], suggest logistic regression is able to predict an individual outcome.

The objective of this study is to develop a BDPM based on the condition of building rating. The target variable is a condition rating of the building and the set of predictors which is independent variables are the factors and building defects. The true relationship between dependent and independent variables can be approximated by regression model. Yong and Sulieman [3], noted that the ratio of sample size to independent variable used in the study is important. The sample size used in this study is 303, the independent variables for factor contributing to the building defect is 21. A minimum requirement for sample size is to have fives time more than the number of independent variable [58].

5.8.7 *Qualitative Analysis*

Qualitative data used were transcript of individual interview and focus group, field notes from observation, audio/video recording and etc. [59]. Miles et al. [60], suggested that qualitative analysis consists of data reduction, data display, and conclusions. Qualitative data analysis is a range of process and procedures transformed the qualitative data collected in a form of explanation, interpretation of situation according to the research [59].

Miles et al. [60], stated that data reduction is referring to the process of selecting, focusing, simplifying, abstracting or transform of the data. In this study, the data reduction was used to produce the interview summary. The results of the interview will be divided into 3 sections which is contribution factors, defects and condition rating according to the objectives of the research.

Data display is the second element in [60], model of qualitative data analysis. The number of participants and responses were recorded with coding including the elaboration of the reason in a related format. Finally, the conclusion of the analysis was described in writing. In this study, the results from interview session were compared with quantitative results.

5.8.7.1 Content Analysis

Content analysis is widely used, and it is used to interpret the meaning of the content of text data [61]. Kondracki et al. [62], noted that text data such as verbal data, print or electronic form obtained from articles, books, manuals, interviews, observations, newspaper, and others. There are three step that should be followed in preparing the content analysis [33].

Firstly, in this study the content of all reports has been analysed to identify factors contributing to the building defects and assessing the extent of defects of the building. Secondly, the information gathered was from the document based on the criteria set up in the sampling method. Thirdly, the factors identified have been analysed using relative index (RII) to determine the relative importance of the factors causing the building defects. The results were then combined with condition survey results to come out with the conclusion of the study. The content analysis from recording of interview was used to validate the overall results.

The analysis must be systematic, generalizable and quantifiable included the following [33]:

(i) Type of study;
(ii) Problem or question of concern;
(iii) Document analysed;
(iv) Document analysis process;
(v) Coding process and categorical;
(vi) Patterns that emerged;
(vii) Assertions and implications.

5.9 Data Interpretation

Interpretation of the result is the process of making sense of numerical data that has been collected and analysed. The researcher should describe clearly their interpretation of the result and in a way other researcher can compare them with their own results. The results need to be interpreted in an objective and critical way based on the research question and research objectives.

In this study, the analysis results were discussed from the condition assessment, archived documentation analysis and semi-structured analysis. The results gathered from condition survey and archive documentation are merged and compared with previous research in literature and theory. Then the interview results are compared to validate the overall results in the study. In this study, the process of document analysis as shown in Fig. 5.6.

5.10 Legitimation/Validity and Reliability

Legitimation demonstrates the reflexivity which helped improved validity [48, 63]. This refers to whether the findings are believable and applicable to another context respectively. Ali [12], explained that validity in research is the strength of research conclusion, inferences or propositions. Validity is the ability of an instrument to measure what it is supposed to measure. In this study, the theoretical framework also validate by the interview results is used to compare and validated the overall results.

5.11 Conclusion

Finally, the findings and analysis are brought to the conclusion of the research. Recommendation and suggestion are produced in the conclusion. Testing and validation were carried out for the proposed BDPM to prove it is beneficial.

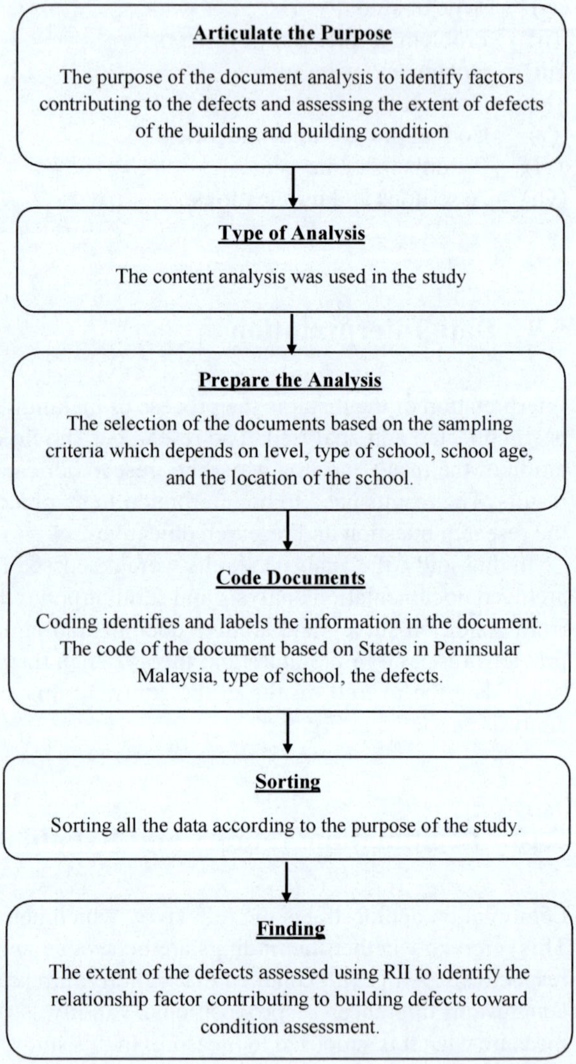

Fig. 5.6 The process of document analysis [33]

References

1. Creswell, J. W., & Clark, P. V. L. (2007). *Designing and conducting mixed methods research*. SAGE Publication, Inc.
2. Graziano, A. M., & Raulin, M. L. (2010). *Research methods: A process of inquiry* (7th ed.). Pearson Education Inc.
3. Yong, C. Y., & Sulieman, M. Z. (2015). Assessment of building maintenance management practice and occupant satisfaction of school buildings in Perak, Malaysia,. *Jurnal Teknologi*, 1–6.
4. Andriotis, K. (2011). A comparative study of visitors to urban, coastal and rural areas—evidence from the Island of Crete. *European Journal of Tourism Research, 4*(2), 93–108.

5. Johnson, R. B., & Onwuegbuzie, A. J. (2004). Mixed methods research: A research paradigm whose time has come. *Educational Researcher, 33*, 14–26.
6. Sani, S. I., Mohammed, A. H., Misnan, M. S., & Awang, M. (2012). Determinant factors in development of maintenance culture in managing public asset and facilities. *Procedia Social and Behavioral Sciences*, 827–832.
7. Watt, D. (1999). *Building pathology: Principles and practice*. Blackwell Science.
8. Clark, P. (2010). The adoption and practice of mixed methods: U.S trends in federally funded health-related research. SAGE Publication.
9. Johnson, R. B., & Turner, L. A. (2003). Data collection strategies in mixed methods research. In *Handbook of mixed methods in social and behavioral research.*
10. Mwaniki, C. N, & Mue, J. S. (2015). Mixed method research: The hidden cracks of the triangulation design, Mount Meru University Research Unit (pp. 46–67).
11. Yeasmin, S., & Rahman, K. F. (2012). Triangulation' research method as the tool of social science research. *BUP Journal, 1*, 154–163.
12. Ali, A. S. (2008). Integrative mechanisms in the design process of building refurbishment projects. PhD Unpublished Thesis, Universiti Teknologi MARA, Shah Alam.
13. Ahluwalia, S. S. (2008). *A framework for efficient condition assessment of the building infrastructure*. University of Waterloo.
14. Fox, J. (1997). *Applied regression analysis, linear models, and related methods* (p. 597). Sage Publications.
15. Wittink, M. N., Barg, F. K., & Gallo, J. J. (2006). Unwritten rules of talking to doctors about depression: Integrating qualitative and quantitative methods. *Annals of Family Medicine, 4*(4), 302–309.
16. Baird, G., Gray, J., Issacs, N., Kemohan, D., & Mcindoe, G. (1996). *Building evaluation techniques*. McGraw-Hill.
17. PWD. (2011). *Directory of federal government premises registration*. Malaysia Public Work Department.
18. Cobbinah, P. J. (2010). Maintenance of building of public institutions in Ghana. Case study of selected institutions in the Ashanti region of Ghana. Unpublished master's thesis, Kwame Nkrumah University of Science and Technology, Kumasi.
19. PWD. (2011). *Sistem Pengurusan Aset Tak Alih (MySPATA)*. Malaysia Public Work Department.
20. Rusli, E. A., Mydin, M. A., & Abdullah, S. (2014). Study on significant mechanisms of maintenance culture among public building users in Malaysia. In *SHS Web of Conferences* (Vol. 11, pp. 1–6). EDP Sciences.
21. Mahli, M., Che-Ani, A. I., Abd-Razak, M. Z., Tawil, N. M., & Yahaya, H. (2012). School age and building defects: Analysis using condition survey protocol (CSP) 1 matrix. *International Journal of Civil, Environmental, Structural, Construction and Architectural Engineering, 6*(7), 56–58.
22. National Centre for Education Statistic (NCES) 2000. U.S Department of Education.
23. IKRAM Report of Overall Trend Assessment & Analysis Report. (2011). Government of Malaysia.
24. Krejcie, R. V., & Morgan, D. W. (1970). Determining sample size for research activities. *Educational and Psychological Measurement, 30*, 607–610.
25. Chong, W. K., & Low, S. P. (2006). Assessment of defects at construction and occupancy stages. *Journal of Performance of Constructed Facilities*, 213–221.
26. Yunus, F. W., & Ali, Z. M. (2013). Attitude towards learning chemistry among secondary school students in Malaysia. *Journal of Asian Behavioural Studies, 3*(11), 1–11.
27. Lewis, B. T., & Payant, R. P. (2000) *Facility inspection field manual: A complete condition assessment guide* (McGraw-Hill).
28. Baker, S. E., & Edward, R. (2012). How many qualitative interviews is enough. National Centre for Research Methods Review Paper (pp. 1–42).
29. PWD. (2013). *Building condition inspection for existing building*. Malaysia Public Work Department.

30. Silva, C. N. (2008). Review: Catherine Marshall & Gretchen B. Rossman (2006). Designing qualitative research [20 paragraphs]. *Forum Qualitative Sozialforschung / Forum: Qualitative Social Research, 9*(3), Art. 13, http://nbn-resolving.de/urn:nbn:de:0114-fqs0803137
31. Othman, N. L, Jaafar, M., Wan-Harun, W. M., & Ibrahim, F. (2015). A case study on moisture problems and building defects. *Procedia-Social and Behavioral Sciences*, 27–36. Seoul, S.Korea.
32. Megha, D., & Rajiv, B. (2013). *Critical Cases of Delay in Residential Construction Projects: Case Study of Central Gujarat Region of India, 4*(4), 762–766.
33. Pershing, J. L. (2002). Using document analysis in analyzing and evaluating performance. *Performance Improvement, 41*, 36–42. Retrieved from http://www.ispi.org
34. Opdenakker, R. (2006, September). Advantages and disadvantages of four interview techniques in qualitative research. *Forum Qualitative Sozialforschung/Forum: Qualitative Social Research, 7*(4).
35. Wilson, V. (2014). Research methods: Triangulation. *Evidence Based Library and Information Practice, 9*(1), 74–75.
36. Cohen, D., & Crabtree, B. (2006). Qualitative research guidelines project.
37. Stuart, A. L., Pasco, J. A., Jacka, F. N., Brennan, S. L., Berk, M., & Williams, L. J. (2014). Comparison of self-report and structured clinical interview in the identification of depression. *Comprehensive Psychiatry, 55*(4), 866–869.
38. Saunders, M., Lewis, P., and Thornhill, A. (2009). Research method for Business Student fifth edition, Prentice Hall
39. Georgiou, J., Smith, J., & Peter, E. D. (2014). A benchmark study on housing defects in Victoria, Australia. *International Journal of Construction Management*, 37–41.
40. MacLean, L. M., Meyer, M., & Estable, A. (2004). Improving accuracy of transcripts in qualitative research. *Qualitative Health Research, 14*(1), 113–123. https://doi.org/10.1177/1049732303259804
41. Keller, G., Warrack, B., & Bartel, H. (1990). *Statistic for management and economics* (2nd ed.). International Thomson Publishing.
42. Trochim, W. M., Cabrera, D. A., Milstein, B., Gallagher, R. S., & Leischow, S. J. (2006). Practical challenges of systems thinking and modeling in public health. *American journal of public health, 96*(3), 538–546.
43. Park, B., Flowerday, T., & Brünken, R. (2015). Cognitive and affective effect of seductive details in multimedia learning. *Computers in Human Behavior, 44*, 267–278. https://doi.org/10.1016/j.chb.2014.10.061
44. Gravetter, F. J., & Wallnau, L. B. (2000). *Statistics for the behavioral sciences* (5th ed.) Wadsworth—Thomson Learning.
45. Foster, L., & Diamond, I., & Banton, J. (2014). *Beginning statistics: An introduction for social scientists*. SAGE 147390921X, 9781473909212. Length 304 pages.
46. Ghasemi, A., & Zahediasi, S. (2012). Normality tests for statistical analysis: A guide for non-statisticians (pp. 486–489), Spring Publication.
47. Pallant, J. (2007). *SPSS survival manual, a step by step guide to data analysis using SPSS for windows* (3rd ed., pp. 179–200). McGraw Hill.
48. Zou, K. H., Tuncali, K., & Silverman, G. S. (2003). Correlation and simpe linear regression. *Journal of Radiology, 227*(3).
49. Tavakol, M., & Dennick, R. (2011). Making sense of Cronbach's alpha. *International journal of medical education, 2*, 53–55. https://doi.org/10.5116/ijme.4dfb.8dfd
50. Gliem, J. A., & Gliem, R. R. (2003). Calculating, interpreting, and reporting Cronbach's Alpha reliability coefficient for Likert-type scales. In *Midwest Research to Practice Conference in Adult, Continuing, and Community Education* (pp. 82–88).
51. Sekaran, U., & Bougie, R. (2009). *Research methods for business: A skill-building approach* (5th ed.). John Wiley and Sons Inc.
52. Leech, N. L., Barrett, K. C., & Morgan, G. A. (2011). *IBM SPSS for intermediate statistics: Use and Interpretation* (4th ed.). Taylor and Francis Group LLC.

53. Assa, M., & Abdi, E. K. (2012). Selected macroeconomic variables affecting private investment in Malawi.
54. McDonald, J. H. (2008). *Handbook of biological statistic*. Sparky House Publishing Baltimore.
55. Chatterjee, S., & Hadi, A. S. (2012). *Regression analysis by example*. John Willey.
56. Liang, K., & Zeger, S. L. (1993). *Regression analysis for correlated data*. John Hopkins University, Baltimore Maryland.
57. Osborne, J. O., & Waters, E. (2002). Practical assessment research & evaluation. *A Peer-Reviewed Electronic Journal, 8*, 1–5.
58. Coakes, S. J., Steed, L., & Ong, C. (2009). *Analysis without Anguish*. John Wiley & Sons.
59. Nigatu, T. (2009). Qualitative data analysis. Africa Research and Medical Foundation, viewed April, 22, 2015.
60. Miles, M. B., & Huberman, A. M. (1994). *Qualitative data analysis: An expanded sourcebook*. Sage.
61. Hsieh, H. F., & Shannon, S. E. (2005). Three approaches to qualitative content analysis. *Qualitative Health Research, 15*(9), 1277–1288.
62. Kondracki, N. L., Wellman, N. S., & Amundson, D. R. (2002). Content analysis: Review of methods and their applications in nutrition education. *Journal of nutrition education and behavior, 34*(4), 224–230.
63. Johnson, R. B., Onwuegbuzie, A. J., & Turner, L. A. (2007). Toward a definition of mixed methods research. *Journal of Mixed Methods Research, 1*, 112–133.

Chapter 6
Data Analysis and Research Findings

Abstract From the study, twenty-one (21) variables were found to be independent variables toward building condition as dependent variable in the study. The variables were analysed to identify the ranking of the result analysis. The other analysis was conducted such as factor analysis to test the relationship between factors toward building condition. The correlation analysis was performed to identify the significant relationship between the variables. The logistic regression model was produced for prediction purpose. Only one model was acceptable to produce to predict the building condition. Three (3) independent variables significantly improve the model which is poor waterproofing, lack of maintenance; and lack of cleaning. The regression model is useful to predict the condition of the building. The condition of the building will be improved by eliminating or reducing the source of factors which can lead to building defects. The condition assessment finding was obtained through archived documentations and the result were analysed together with the interview results to validate the findings. The results were valid, and the conclusion of the research would be discussed in Chap. 7.

6.1 Introduction

This chapter presents the analysis of data and research findings from the triangulation method. Data were analysed to identify, to analyse, and to explore the relationship between factors which contributes to the building defects and their effects toward building condition ratings and to develop the BDPM based on the building condition. Quantitative data were obtained from field work condition survey and qualitative data from archived documentation and semi-structured interview. The data were then combined and merged to accomplish the research objectives.

This study is designed to establish the significant relationship between the factors and building condition. Twenty-nine (29) variables were identified from the research objective and building condition rating as dependent variables. Each result findings are explained to answer the objectives of the research as shown in Table 6.1. The result findings then analysed using the statistical analysis software Statistical Package for

S. Yacob et al., *Managing Building Deterioration*, Management in the Built Environment,
https://doi.org/10.1007/978-981-16-5860-0_6

Table 6.1 Analysis of results to research objectives

No	Type of SPSS analysis result	Research objectives
1	(a) Descriptive (b) Reliability (c) Content analysis	(1) To identify the factors contribute to the buildings defects (2) To analyse current condition of the buildings through condition survey
2	(a) Correlation (b) Content analysis	(3) To establish significant relationship between the factors and building condition
3	Logistic regression	(4) To develop a building deterioration prediction model (BDPM) based on condition of the buildings

Social Science (SPSS). The twenty-nine (29) variables which form the central theme of the study were tested using both descriptive and inferential statistics. Findings results will be further analysed using the following techniques/tests.

6.2 Quantitative Analysis

6.2.1 Descriptive Analysis Result

In this study, quantitative analysis refers to the determination of how much of a given component in the samples of the 303 school buildings.

6.2.1.1 Type of Schools

Figure 6.1 shows the total number of schools analysed in the study which comprised of primary and secondary schools. The total number for primary schools building is 175 while for secondary schools building is 128 with an overall total of N = 303.

Fig. 6.1 Type of schools

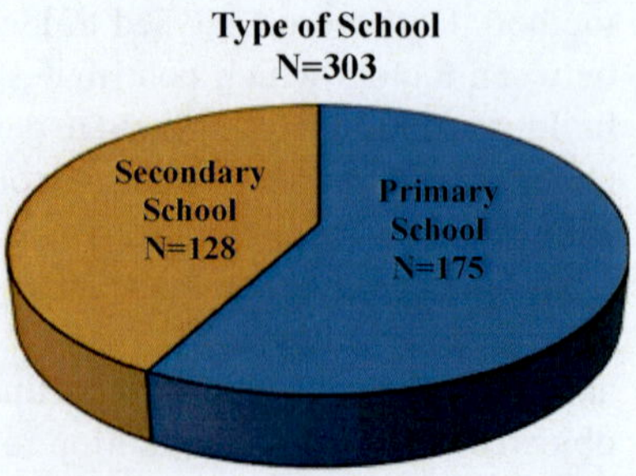

6.2.1.2 The Number of Schools According to School Age

Based on the data obtained through the content analysis of the reports, the age of the schools is between 9 and 20 years according to the sampling criteria set up for this study. The analysis of the school is for the year 1995–2006. Figure 6.2 shows that the total number of schools according to the year and building age.

Majority of the schools analysed for age 12 years which was built in 2003 with a total number of 54. Then, it is followed by 47 buildings for age 20 years built in 1995. For school building built in 1996 is 10 in number with age of 19 years, 7 schools built in year 1997, 11 schools for year 1998, 14 schools for year 1999, 25 schools for year 2000, 43 schools for year 2001, 35 schools for year 2002, 32 schools for year 2004, 18 schools for year 2005, and 7 schools for year 2005.

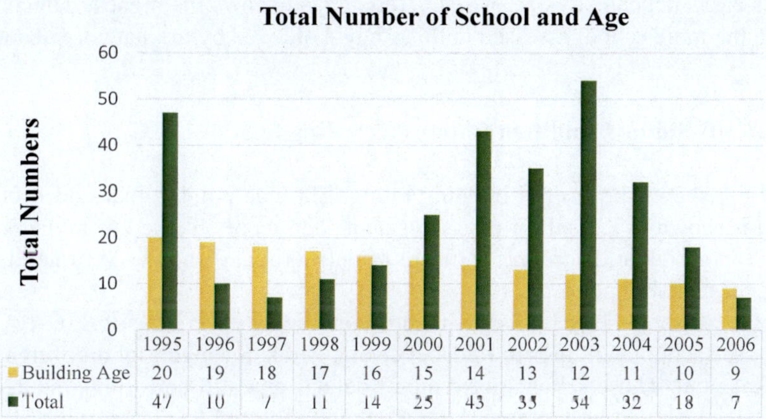

Fig. 6.2 Total number of schools according to building age

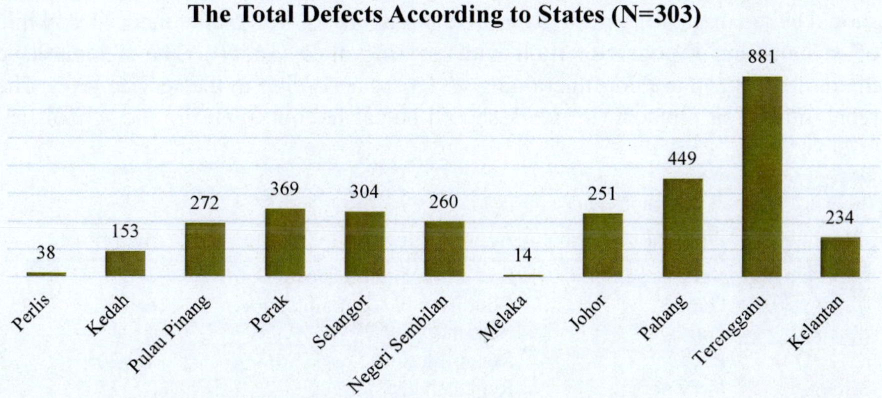

Fig. 6.3 Total defects according to states

6.2.1.3 Total Defects According to State

The South area comprises the state of Johore and Malacca. The total number of defects found for the state of Johore is 251 and Malacca is 14 defects. In the West area, the state of Perak has 369 defects, the state of Selangor has 304 defects and the state of Negeri Sembilan has 260 defects. The number of schools are different in each state due to and depending to the size of the state. The state of Terengganu shows the highest defects found due to more schools assessed in this study, followed by the state of Pahang.

South area comprised the state of Johor and Melaka. The total number of defects found for the state of Johore is 251 and Melaka with only 14 defects. For West area, the state of Perak with 369 defects, the state of Selangor with 304 defects and the state of Negeri Sembilan with 260 defects. The number of schools are different in each state is depends to size of state. Terengganu shows the highest defects found due to the more school assessed in this study. Followed by the state of Pahang.

6.2.1.4 Building Condition Rating According to State

Table 6.2 shows the overall building rating. The five-point colour-coded building rating is shown as a result of the assessment. The score obtained from the building condition schedule determine the level of defects found such as very good, good, fair, poor and very poor.

Table 6.3 shows 16 schools are in good condition with a rating of B (6–10) which comprised school buildings in Kedah, Penang, Perak, Selangor, Johore, Terengganu, and Kelantan. Those schools have minor defects and still performing as intended. The defects should be monitored, repaired and replaced to prevent serious defect. The results also show 221 school in average condition with a rating of C (11–15). Terengganu is the highest state that has school buildings with condition C. The schools have major defects, however, still can function with supervision. The defects should be repaired or replaced to ensure the schools are well maintained. Meanwhile 65 schools are in poor condition with a rating of D (16–20). The schools have major defects and are not functioning as agreed according to the service level. The rehabilitation or replacement work should be carried out to ensure the schools are

Table 6.2 Overall condition rating

Rating	Physical condition	Action matrix	Score
A	Very good	Preventive maintenance	1–5
B	Good	Condition based maintenance	6–10
C	Fair	Repairs	11–15
D	Poor	Rehabilitation	16–20
E	Very poor	Replacement	21–25

Source PWD JKR 21602–0004-13 (2013)

Table 6.3 Overall building condition of school according to the states

No	State	Condition rating				
		A	B	C	D	E
		Very good	Good	Fair	Poor	Very poor
		0–5	6–10	11–15	16–20	21–25
1	Perlis	0	0	1	2	1
2	Kedah	0	2	7	7	0
3	Penang	0	2	18	10	0
4	Perak	0	3	21	8	0
5	Selangor	0	1	20	9	0
6	Negeri Sembilan	0	0	14	7	0
7	Melaka	0	0	2	0	0
8	Johore	0	1	20	9	0
9	Pahang	0	0	33	8	0
10	Terengganu	0	3	64	4	0
11	Kelantan	0	4	21	1	0
	Total	0	16	221	65	1

functioning. Only 1 school is in a very poor condition with the rating of E (21–25). This means that the school is in a very critical, not functioning, risky to safety and health and major replacement is need.

Figure 6.4 shows the overall condition rating of the school building and detailed condition rating for each state. The building condition C with a fair condition is the highest rank with a total of 221 schools or 72.67%. Second highest rank is schools with critical condition with 65 schools or 21.67%. A total of 16 schools or 5.33% is in a good condition. Only 1 school or 1% was in very poor condition.

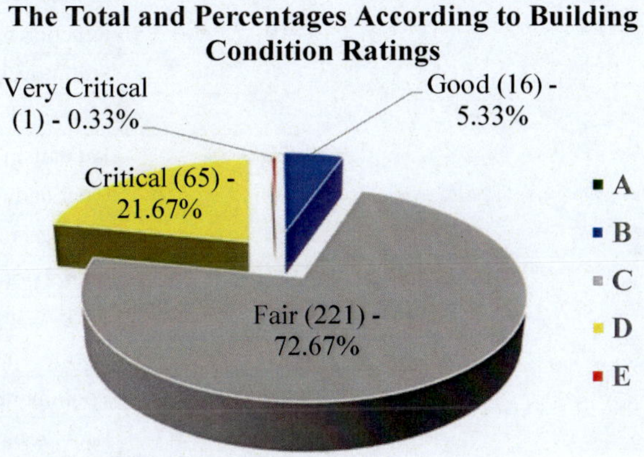

Fig. 6.4 The total and percentage of condition ratings

6.2.1.5 The Factors Contribute to the Building Defects

In this study, the reliability analysis was conducted to determine the reliability of data collected. A total of 29 factors were adopted in this study. These factors were used in the data collection for the study. The factors affecting the defects in the buildings scored on a five-point Likert Scale based on their significant. Table 6.4 shows the 29 factors that causes building defects.

Table 6.4 List of factors

No	Factors
1	Faulty design
2	Poor detailing
3	Faulty drawing
4	Not complying with specification
5	Structural
6	Poor waterproofing
7	Improper use of material
8	Lack of maintenance
9	Poor construction
10	Building size
11	Type of building
12	Building orientation
13	Insect attack
14	Biological agent
15	Change of climatic condition
16	Poor ventilation
17	Reaction of thermal agent
18	Excessive moisture
19	Reaction of chemical agent
20	Soil movement impact
21	Misuse by user
22	Vandalism
23	Wear and tear
24	Changes of usage
25	Poor workmanship
26	Lack of supervision
27	Lack of knowledge
28	Poor operation
29	Lack of cleaning

6.2.1.6 Types of Defects

In this study, the various types of defects were also identified. Tables 6.5 and 6.6 shows the defects were found and categorized into two (2) types of defects which is defect under building element and defects under mechanical and electrical component according to PWD [1]. These types of defects were analysed in this study to determine the correlation between defects and building condition rating to predict the factors that contribute to building defects towards building condition rating.

Figure 6.5 demonstrated the total number of building defects found in the study. It was found that Broken is the highest rank with a score of 220 in numbers. In

Table 6.5 Types of building defects

No	Buildings defects
1	Bent
2	Blockage
3	Broken
4	Corrosion
5	Crack
6	Damage
7	Dampness
8	Decay
9	Dirty
10	Discolouration
11	Exposed rebar
12	Leakage
13	Loose
14	Misplaced
15	Misused
16	Mould
17	Overflow
18	Sagging
19	Settlement
20	Spalling
21	Termite attack
22	Torn out
23	Uneven surface
24	Improper installed
25	Vegetation
26	Water ponding
27	Water seepage
28	Water stain

Table 6.6 Types of mechanical and electrical (M&E) defects

No	M& E defects
1	Exposed wiring
2	Insufficient light
3	Misalign
4	Missing
5	No gas
6	Not functioning
7	Not installed
8	Overloading
9	Short circuit
10	Validity certificate expired

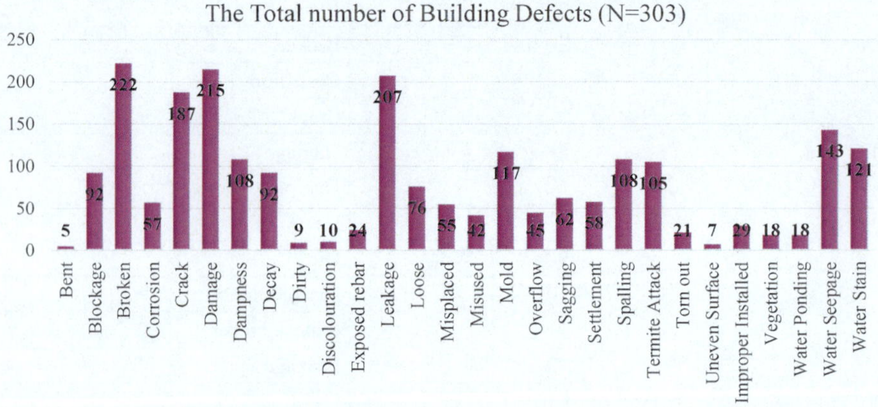

Fig. 6.5 The total number of building defects

this finding, "Broken" refers to roof and floor tiles, door, window, sanitary fitting, floor and walls. The findings supported the statement of Zamri and Abdullah [2] that broken always happened to roof and floor tiles. The study by Roslan et al. [3], also mentioned that broken to roof and floor tiles is one of building defects that occurred in buildings. Damage to building elements ranked in second with a value of 214 number of finding. According to Macabuag et al. [4], damage to school is due to insufficient transverse reinforcement to column.

In this study a lot of damages is due to door, window, sanitary fittings, wall and floor tiles which cannot perform as it is and as required. This result is supported by Noy [5], that common damages with window and door is for ironmongery such as loose hinges, worn bolts, shrinkages in the stiles and damage with fastening. Meanwhile the research by Che Ani et al. [6], mentioned that damage to wall, especially to tiles is due to unacceptable under the current standard.

Leakage is ranked in third with a value of 204 in number. Leakage in this study refers to leakage to roof, ceiling, piping services, gutters, floor and wall. According to Choudhry et al. [7], water leaks is the main culprit and can cause significant structural damages, inconvenience, irrigation issue and trouble. This result findings were supported by Kudder and Erdly [8], that leakage in building was significantly due to water penetration to solid concrete elements, joints of opening, through plastering, weep holes, drainage, gutter, inner wall and through cracking of wall and floor element.

Figure 6.6 demonstrated the total number of mechanical and electrical defects found in the study. It was found that not functioning as the highest rank with a score of 278 in number. In this study, not functioning referred to firefighting appliances, switch, sockets, distribution board, and other mechanical and electrical fittings. According to research by National Research Council [9], not functioning to the appliances or fitting not served as acceptance standard due to lightning, damages, power loss, and power utility and also failed to calibrate the mechanical and electrical equipment.

Second highest rank with a score of 230 in number is validity certificate expired. This is due to certificate expired for fire extinguisher for all school buildings in this study. According to Fire Services Regulation [10], fire extinguishers should be checked, and the certificate should be renewed annually to ensure proper maintenance and safety during occupation of the building.

The result shows that the school building failed to renew the certificate for extinguisher according to the regulation. This is in line with the report by New Strait Times (2016), which found that the number of fire cases involving schools greatly increased due to the ignorance of owner of the building to check the firefighting equipment is well maintained and usable.

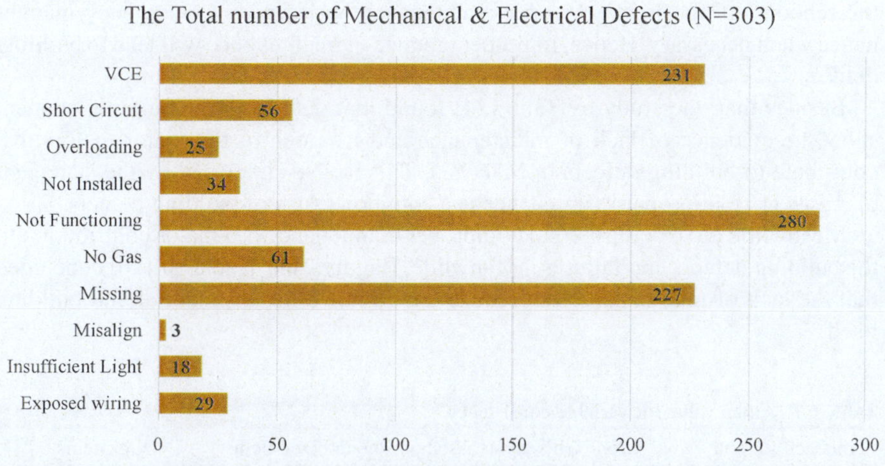

Fig. 6.6 Total number of mechanical and electrical defects

6.3 Reliability

From the analysis, it was found that from 29 number of factors analysed, only 21 factors were ranked according to the mean reading for most significant factor that contribute to defects. For the other factors such as poor detailing, faulty drawing, building size, building types, building orientation, lack of knowledge, poor ventilation and poor operation has zero variance and is thus removed from the scale.

The first data was run for reliability analysis for internal consistency measure. Table 6.7 shows the alpha coefficient for the 21 variables is 0.713 and have acceptable internal consistency. According to the research by Mohsen [11], the acceptance value for alpha ranging from 0.70 to 0.95.

Table 6.8 shows the rank of factors that contributes to the building defects. The result shows a lack of supervision as the most significant factor with mean value of 4.5809. According to Waziri [12], as quoted in the studies by Assaf, Al-Hammad, and Al - Shihah [13], Okuntade [14], lack of supervision will increase the construction defects and lead to high maintenance during operation.

Lack of supervision was identified to have a significant relationship to increase rate of building failures or defects [15]. These findings supported the findings of Ahzahar et al. [16], lack of supervision has a major influence on the quality of the building which can cause building defects or causes of rework. This always has an impact toward building condition. Meanwhile according to Ali et al. [17], lack of supervision caused the quality problem which can lead to the defects in the building.

Lack of maintenance factor shows the second highest rank with a mean value of 4.5710 from all factors that contribute to the defect. The finding confirmed maintenance will cause more damages and defects if left unattended [18]. The finding is also consistent with the findings of past studies by Wardhana [19], Chew [20], Constantinos [21], Md Ulang [22]. Research by Talib [24], reported that most of the schools do not practice proper maintenance but carry out emergency maintenance when necessary. Hence, improper practice of maintenance will lead to building defects.

Besides that, the study by Talib [24] found that defects to the building element show the evidence of lack of maintenance and it is one of the major factors that contribute to building defects in Malaysia. This is also supported by the statement of Agwu [15], improper maintenance is a common cause of building defects.

Meanwhile [25], confirmed that improper maintenance is a major contribution to the building defects and failures. Meanwhile, Bagdiya and Wadalkar [26] concluded that the lack of maintenance can reduce the effective life of material and building itself.

Table 6.7 Alpha value for variable consistency

Cronbach's alpha	Cronbach's alpha based on standardized items	N of items
0.713	0.712	21

Table 6.8 Ranking of factors that contribute to building defects according to mean

Rank	Factors which contribute to building defects	Mean (N = 303)	Std. deviation
1	Lack of supervision	4.5809	0.88357
2	Lack of maintenance	4.5710	0.70058
3	Vandalism	3.8680	1.20008
4	Improper use of material	3.4422	1.16619
5	poor workmanship	3.3696	0.71581
6	Wear and tear	3.2013	1.03674
7	Poor construction	3.1914	1.25538
8	Excessive moisture	3.0462	1.21894
9	Poor waterproofing	2.7228	1.09317
10	Change of climatic condition	2.7162	1.08474
11	Misuse by user	2.6799	1.60578
12	Biological agent	2.5941	1.11817
13	Insect attack	2.3861	1.14953
14	Soil movement impact	2.0594	1.70878
15	Lack of cleaning	1.9175	1.18340
16	Reaction of chemical agent	1.7492	1.11126
17	Faulty design	1.6271	1.15210
18	Changes of usage	1.6139	1.03550
19	Reaction of thermal agent	1.4587	0.93715
20	Structural	1.4587	0.94770
21	Not complying with specification	1.4455	0.93294

The third factor is vandalism with a mean value of 3.8680. The findings verified that vandalism is a problem facing educators and it will increase the cost of repair to the building element or property [27].

Study by De Wet [28], suggests it is evident from the literature that the defects caused by vandals such as broken window, indoor damage especially grafifi on wall, and broken fitting in the bathroom. Vandalism refer to willful or malicious damage to school ground, building or furnishing and equipment [29]. Lack of care or attention can lead to vandalism which destroys school property and the damage to building element. Vandalism is the actions that can damage the school facilities and infrastructure [23].

The fourth factor is improper use of material with a mean value of 3.4422. One of improper design factors is a selection of materials [30]. According to Chew [20], selection of material is most important to control building defects. The other factors with less values of mean are poor workmanship with value 3.3696. Construction project suffered from the low quality of workmanship and most significant factor that contribute to workmanship is a lack of experience [17]. Poor workmanship is one of the factors that lead to building defects [18]. For wear and tear factors with a mean value of 3.2013, poor constructions with mean value of 3.1914, excessive moisture with mean value of 3.0462.

The others factors with mean values below 3.000 are poor waterproofing with mean value of 2.7228, change of climatic condition with mean value of 2.7162, misuse by user factor with mean value of 2.6799, reaction to biological agent with value mean of 2.5941, insect attack with mean value of 2.3861 and soil movement impact with mean value of 2.0594. Meanwhile, factors which have the mean value of less than 2.000 are lack of cleaning, reaction to chemical agent, faulty design, changes of usage, reaction thermal agent, structural and not complying with specification.

6.4 Correlation Analysis

6.4.1 To Establish a Significant Relationship Between Factors and Building Condition

This research is designed to establish a significant relationship between the factors and building condition. Twenty-one (21) significant factors were ranked according to the mean reading were run for the correlation analysis. The correlation test using Spearman rank correlation coefficient was used to check the significance relationship between factors and building condition. Correlation is a statistical technique that is useful to determine if there exists a linear relationship between two variables. It is also a useful tool to show how strong or significant is the relationship between the variables.

The correlation coefficient or "r" is normally used to measure the strength of the relationship and reported in a value between -1 and $+1$. A correlation coefficient, r value close to -1 or $+1$ showed a strong relationship between the two variables. The r value close to 0 indicates little or no relationship between the two variables. Furthermore, when the correlation is positive or "r" value positive, it shows that the two variables are increasing in the same direction and vice versa for negative r value or inverse correlation.

Further analysis was done to determine the relationship between the overall building conditions rating and the various factors that contribute to the building deterioration. For a model with dependent variable condition rating, the following variables are constants or have missing correlations: poor detailing, faulty drawing, building size, building types, building orientation, poor ventilation, lack of knowledge and poor operation. They will be deleted from the analysis and only twenty-one (21) of the variables will be used as a sample in the analysis.

Table 6.9 shows the ranking of the relationship between defects factors and building condition. From the results, it was found that only 6 factors that contribute a significant relationship with significant value $p < 0.05$ with building condition rate of the school building. A correlation coefficient, r value with $+1$ showed a strong relationship between the two variables. There appears to be a positive correlation

Table 6.9 The relationship between factors that contribute to building defects and building condition

No	Factors that contribute to building defects	Building condition rate
1	Lack of maintenance	0.311**
2	**Vandalism**	**0.273****
3	**Poor waterproofing**	**0.253****
4	**Lack of supervision**	**0.181****
5	**Lack of cleaning**	**0.158****
6	**Misuse by user**	**0.147***
7	Improper use of material	0.106
8	Insect attack	0.066
9	Faulty design	0.052
10	Wear and tear	0.036
11	Poor workmanship	0.035
12	Not complying with specification	0.030
13	Reaction of thermal agent	0.023
14	Excessive moisture	0.021
15	Poor construction	0.015
16	Changes of usage	−0.001
17	Soil movement impact	−0.007
18	Structural	−0.077
19	Change of climatic condition	−0.083
20	Reaction of chemical agent	−0.095
21	Biological agent	−0.004

*Correlation is significant at the 0.05 level (2-tailed)
**Correlation is significant at the 0.01 level (2-tailed)

between two variables. Six (6) factors have positive correlation relationship which is lack of maintenance, vandalism, poor waterproofing, lack of supervision, lack of cleaning and misuse by user.

Although the other factors are expected to have a relationship with building condition, but according to the PWD inspection report, not all the factors contribute the same defects for each school assessed and not all the factors found affected the building condition of the school. The defect is generated from the various factors based on the type of defects found at the element of the building. They vary depending on the types of school, location of the school and the size and number of defects found.

6.4.1.1 Relationship of Lack of Maintenance to Building Condition

Lack of maintenance shows a value of 0.311 which is highest significant relationship to the building condition. The important factor which leads to building defect is maintenance [24]. Maintenance has positive correlation which means lack of maintenance is the highest factor that leads to building defects. The condition of the building and highly deteriorate. If the maintenance work is carried out comprehensively, the condition rating is also increased to better condition.

*Correlation is significant at the 0.05 level (2-tailed)

**Correlation is significant at the 0.01 level (2-tailed)

According to Yacob [31], maintenance of school building are usually neglected and influenced the physical condition of the buildings. Ahzahar et al. [16], stated that maintenance plays a major role in preventing building defects and pointed that the building which is neglect for maintenance may fall into several defects and lead to the structural failures, future financial burden, pose legal issues and etc. It is important that buildings continue to be properly maintained to ensure that they can function as efficiency and effectively as possible [16].

6.4.1.2 Relationship of Vandalism to Building Condition

The third important factors are vandalism factors with a value of 0.273. Vandalism at school potentially could damage the system, and the element of the building [23]. According to Chong and Low [32], softer building material often invites vandalism and lead to defects such as cracks, holes, and etc. which affect the building condition. It can be concluded that without proper monitoring from school management on the vandalism issues will affect the building condition of a school building. This statement is supported by Morton [33], that vandalism can decrease the condition and value of the buildings, resulted in repeated maintenance and incur repair costs.

6.4.1.3 Relationship of Poor Waterproofing to Building Condition

Poor waterproofing factors have a significant relationship to the building condition with a value of 0.253. According to Suffian [18], one of the common problems of building defect in the building is due to poor waterproofing factors which lead to the leakage, cracks, soil settlement, and wall finishes problem. Waterproofing factors will cause defects such as water leak, tilt, algae/fungus attack, mould, and mildew growth to the element of the building. The problem involving waterproofing always occurred at basement, flat roof, floor at toilet and walls. This supported statement by Jaafar and Othman [30], moisture problem leads to failure of the building. Due to this problem, the condition of the building will deteriorate and will increase the cost of maintenance. The condition of building will be better when the water that penetrate into building decrease by proper construction, installation, repair work related to avoid water to penetrate to the building.

6.4.1.4 Relationship of Lack of Supervision to Building Condition

Lack of supervision to building condition rating has significant relationship with a value of 0.181. According to Ali et al. [17], lack of supervision in term of monitoring work during construction and maintenance will lead to poor workmanship and affect the building condition. Ahzahar et al. [16], site supervision has a major influence on the overall quality of the building and also reported that supervisors have an important role in minimizing the defects in construction. Poor workmanship is also due to negligence and inadequate quality control during the execution of construction work either during or post construction especially during rectifying, replacement or maintenance work. Without proper monitoring on the supervision work will lead to the building defects and affect the condition rating of the buildings.

6.4.1.5 Relationship of Lack of Cleaning to Building Condition

The cleaning factors have a negative correlation with building conditions with a value of 0.158. The school building has significant serious environment problem due to the cleanliness [34]. In this study, the cleaning factors contribute to the types of defects such as blockage of sanitary, water ponding at floor, blockage drain, dirty wall and floor and unhygienic area. This supported the statement of Sreenivasan et al. [35], that unhygienic area will cause safety and occupational health hazards besides being a source of defects. The condition of the building will increase to a better condition when the users are aware to clean, use and operate their buildings accordingly with a proper maintenance. Lack of cleaning will lower the value and the condition of the building.

6.4.1.6 Relationship of Misuse by User to Building Condition

Misuse by user shows score rank with a 0.147. The study by Annila et al. [36], revealed the air conditioning in building have a significant effect to moisture and mould damages. This is due to the high use of air conditioning system by making mistake during the operation of the building. For example, very high temperature in room can caused the condensation outside the room or internal humidity [37] and could lead to the mould damages. In this study, the wrong use of building component can cause defects to the building, for example, broken to sanitary fitting. This statement is supported by Hong [38] damaged sanitary fittings such as water cistern, sinks, washbasin, proven to have been damaged by misuse, mishandling by the user. With a proper awareness and use of the building by the user will improved the condition of the building resulting in positive effect to the user itself, lower future maintenance with fewer damages [39].

6.4.2 To Establish a Significant Relationship Between Number of Defects and Building Condition

In this study, the various defects found were analysed toward building condition rating. According to the correlation result, the building, mechanical and electrical (M & E) defects was found to be significantly correlated with building condition rating as shown in Table 6.10.

The result clearly shows that there is correlation between buildings and M & E defects toward condition ratings. Table 6.10 shows that building and M&E defects have a significant relationship between building condition rating with a score of 0.468 and 0.280. The findings demonstrated that building defects correlated to the building condition. When more defects are found in the building, the conditions of building also deteriorate higher which can lower the value of the building. The condition of the building affected due to the total number of defects found in the building and type of defects have a significant relationship with the condition of the building component [40]. The statement is proven by the analysis result.

6.5 Regression Analysis

In this study, a binary logistic regression analysis was conducted to predict the factors that contribute to the building defects for 303 school buildings using building condition as predictors. The logistic regression analysis for building condition rating was run based on significant correlation between loading factors as its predictor which consisted of 21 factors that contribute to the building defects. In the analysis, the dependent variables were re- coded into two categories which is code 1 for "functioning with minor defects" and code 2 for "functioning with major defects".

Result found that three (3) independent variables significantly improve the model as predictors which are:

(i) Poor waterproofing (PW)
(ii) Lack of maintenance (LOM)
(iii) Lack of cleaning (LOC).

In this study, forward stepwise method was used, and SPSS produced three (3) steps to include the predictors that significantly contributed to the logistic regression

Table 6.10 Correlation between building, M&E defects and building condition rating

	Building condition rating
Building defects	0.468**
M & E defects	0.280**

*Correlation is significant at the 0.05 level (2-tailed)
**Correlation is significant at the 0.01 level (2-tailed)

Table 6.11 Variables in the equation

		B	S.E	Wald	df	Sig	Exp (B)	95% C.I. for EXP(B)	
								lower	upper
Step 1[a]	Lack of maintenance	0.861	0.257	11.23	1	0.001	2.365	1.429	3.912
	Constant	−0.842	1.078	0.611	1	0.435	0.431		
Step 2[b]	Poor waterproofing	0.545	0.245	4.952	1	0.026	1.725	1.067	2.787
	Lack of maintenance	0.930	0.274	11.54	1	0.001	2.535	1.482	4.336
	Constant	−2.315	1.323	3.061	1	0.080	0.099		
Step 3[c]	Poor waterproofing	0.658	0.262	6.294	1	0.012	1.932	1.155	3.231
	Lack of maintenance	0.755	0.288	6.893	1	0.009	2.129	1.211	3.741
	Lack of cleaning	0.445	0.213	4.388	1	0.036	1.561	1.029	2.368
	Constant	−3.479	1.484	5.496	1	0.019	0.031		

[a]Variable(s) entered on step 1: Poor waterproofing
[b]Variable(s) entered on step 2: Lack of maintenance
[c]Variable(s) entered on step 3: Lack of cleaning

model. Step 1 revealed first independent variable significantly predicting the probability of chances building condition rating with $X^2 = 9.760$, $p < 0.05$. Step 2 shows two (2) variables significantly predicting whether or not functioning has occurred ($X^2 = 16.284$, $p < 0.001$). Then, step 3 computed the third variable with $X^2 = 20.394$, $p < 0.001$. In this study also Hosmer–Lemeshow (H–L) goodness of fit is 0.945 shown in step 1, is more than 0.05 and the model fit the data. Meanwhile step 2 shows H–L goodness of fit is 0.846 and step 3 shows goodness of fit is 0.578.

The entire model adequately fit the data. Thus, all model passes the test since the p-value > 0.05 [41]. In this study model in step 3 with 0.578 which means that it is not statistically significant and therefore the model is good fit. The model fails to reject the null hypothesis that there is no difference between observed and model predicted values.

The model explained between 6.50% (Cox & Snell R Square) and 19.2% (Nagelkerke R Square) of the variance in the dependent variable is explained by three (3) predictors in the model with overall cases of 94.7% (0% for functioning with minor defects and 100% for functioning major defects). Independent variables for i.e. waterproofing (OR = 1.932, 95% CI: 1.155–3.231), lack of maintenance (OR = 2.129, 95% CI: 1.211–3.741), and lack of cleaning (OR = 1.561, 95% CI: 1.029–2.368) made a unique statistically significant contribution to the model (X^2 (3) = 20.39, $p < 0.001$).

Table 6.11 also shows that the predictors measuring the poor waterproofing (B = 0.658), lack of maintenance (B = 0.755), and lack of cleaning (B = 0.445) have positive coefficients indicating that higher defects for poor waterproofing, lack of maintenance or higher lack of cleaning occurred at building are associated with a greater likelihood of building condition. According to the regression model produced, there are three (3) factors summarised that the significant predictors of building condition. The factors are poor waterproofing, lack maintenance, and lack of cleaning. In this case also three predictor variables contributed significantly to the prediction. Independent variables i.e. poor waterproofing ($p = 0.012$, $p < 0.05$), lack of maintenance ($p = 0.009$, $p < 0.05$), and lack of cleaning ($p = 0.036$, $p < 0.05$).

The logistic regression equation with actual values of the regression coefficients is produced as follows:

$$\text{The Equation}: Z = -3.479 + .658(\text{PW}) + .755(\text{LOM}) + .445(\text{LOC})$$

This shows that increasing defects due to poor waterproofing, lack of maintenance and lack of cleaning increases the log odds of the condition of the building (become worst). The odds ratio of each independent is the ratio of relative importance of the independent variables in terms of effect on the dependent variable's odds. Exp(B) values indicates that when the level of degree of defects increase due to poor waterproofing factor, the odds ratio is 2 times as likely to increase the building condition become worst.

For maintenance, the odds ratio is 2.00 times as likely to increase the building condition become worst, Lack of maintenance is 2 times as likely to increase the building condition become worst. Meanwhile lack of cleaning is also 2 times as likely to increase the building condition become worst.

6.6 Qualitative Analysis

6.6.1 Content Analysis for Archived Documentation

PWD Inspection Reports are regarded as the main repositories for lesson learnt documents. The documents in a form of reports consist of backgrounds of the projects, maintenance records, photos and defects of the building.

Content analysis is one of the research techniques used in this study to recreate and valid inferences by interpreting and coding textual material. Duriau et al. [], explained that content analysis is a research method at the intersection of the qualitative and quantitative tradition. Archived documentation were analysed based on background of the schools which consisted of school name, address, district, states, Ministry, department, head of department, GPS coordinate, block number, land use, type, year built, project manager, current maintenance, gross floor area, site area, building officer, end user and name of consultant. The analysis was sorted

Table 6.12 Content analysis for archived documentation

No	Selection criteria	Content analysis explanation
1	Year built	The age of school selected in year-built section which is 1995–2004 with age 11–20 years old
2	States and building code	In this study, the building for each state were recorded and coded based on content analysis findings (a) Kedah—KD (b) Melaka—ML (c) Perlis—PR (d) Terengganu—TG (e) Negeri Sembilan—NS (f) Johor—JR (g) Pahang—PG (h) Kelantan—KB (i) Selangor—SG (j) Penang—PP (k) Perak—PK
3	Type of schools	The types of schools recorded for primary and secondary school
4	Assessment findings	There are 29 factors contributing to the building defects and 38 types of defects found in documents through report finding and photograph of defects

according to the background, States, type of schools and assessment finding as shown in Table 6.12.

In this research, content analysis is used to identify the factors that contributing to the building defects and assessing the extent of the defects of the building to identify the condition of the building. The factors identified in this study have been analysed using relative index (RII) to determine the relative importance of the factor causing the building defects. Appendix F shows the information derived from PWD's report translated to quantitative data to come out with the result of the study.

6.6.2 Semi-structured Interview Findings

Semi-structured interviews were constructed and conducted based on the ranking in correlation analysis result to validate the result analysis. The interview result was expected to agree and support the analysis result of the study. The interviews were recorded, and the discussion was quoted to validate the results. The participants from building surveying discipline were coded as Participant BS1 to BS6, and the participants from civil engineer discipline as Participant E1 to E6.

Table 6.13 shows the result gathered from the semi-structured interviews conducted among 12 PWD professionals. The PWD professionals were either from technical or management with a working experience as a BCA building inspector for more than 3 years. In this study, the interview began with a brief explanation of

Table 6.13 Semi-structured interview results

No	Descriptions	Variables validated	Participant (N)	%	Remarks
1	Factors which have strong significant and affect the overall condition of the building	(a) Lack of maintenance (b) Vandalism (c) Poor waterproofing (d) Lack of supervision (e) Lack of cleaning (f) Misuse by user	N = 12 N = 11 N = 10 N = 11 N = 12 N = 12	94%	The interview result validated the ranking analysis result of strong significant factor toward building condition
2	Factors which have least significant and affect the overall condition of the building	(a) Improper use of material (b) Insect attack (c) Faulty design (d) Wear & tear (e) Poor workmanship (f) Not complying with specification (g) Reaction of thermal agent (h) Excessive moisture (i) Poor construction (j) Change of usage (k) Soil movement impact (l) Structural (m) Change of climatic condition (n) Reaction of chemical agent (o) Reaction of biological agent	N = 11 N = 12 N = 12 N = 12 N = 12 N = 11 N = 12 N = 12 N = 12 N = 12 N = 11 N = 12 N = 12 N = 12 N = 12	98%	The interview result validated the ranking analysis result of least significant factor toward building condition
3	Factors which have zero variance to the overall condition of the building	(a) Poor detailing (b) Faulty drawing (c) Building size (d) Building types (e) Building orientation (f) Lack of knowledge (g) Poor ventilation (h) Poor operation	N = 12 N = 12 N = 12 N = 12 N = 12 N = 12 N = 12 N = 12	100%	The interview result validated the ranking analysis result of zero variance factor toward building condition

(continued)

Table 6.13 (continued)

No	Descriptions	Variables validated	Participant (N)	%	Remarks
4	Theoretical framework	(a) Independent variables (b) Mediating variables (c) Dependant variables	N = 12 N = 12 N = 12	100%	The interview result validated the variables of the study
5	The uses of standard building condition assessment	(a) Method (b) Process	N = 12	100%	The result confirmed the method to rate the extent of the building defects used in the study

the study, aims, requirements and confidentiality. The interviews lasted a minimum of 30 min, depending responses during the interview.

6.6.2.1 Interview Result to Ranking Analysis Result

The semi structured interview was conducted to compare between the findings of ranking analysis result in the study. This focused on the following section as below:

(a) Factors which have strong significant and affect the overall condition of the building;
(b) Factors which have least significant and affect the overall condition of the building;
(c) Factors which have zero variance to the overall condition of the building;
(d) Validation of theoritical framework;
(e) The uses of Standard Building Condition Assessment in the study.

(i) Factors which have strong significant and affect the overall condition of the building

Based on the research, a total of 40 factors were found in the literature which caused the building defect. The factors have been arranged accordingly to group base similarities to be 29 numbers of factors. From the analysis, only 21 factors were analysed to identify the significant relationship between factors and building rating. There are 6 factors which have the most significant correlation between factors and building condition. Twelve (12) participants were involved in the interview and the results are shown in Table 6.14.

Firstly, all participants agreed that lack of maintenance factor have strong significant effect on the condition of the building. Almost of the school buildings lack of maintenance or schedule maintenance which lead to building defects.

In the interview conversation, Participant BS3 mentioned that:

Table 6.14 Investigation on strong significance factors towards the building conditions

Factors which have strong significance effect the building condition	No. of participants confirmed the significance	Reason(s)
(a) Lack of maintenance	12	• Lack of maintenance lead to defects • Almost all school building lack of schedule maintenance • Maintenance not properly carried out by school organization/not right person • Less maintenance lead to building deterioration
(b) Vandalism	11	• Based on cases and location • Almost happened to school at rural areas and due to the student's attitude • A lack of proper school discipline will cause building defects
(c) Poor waterproofing	10	• Issue on the warranty • Aging material which can lead to the leakage at floor especially toilet area, and water penetrated through wall caused by rain, dampness
(d) Lack of supervision	11	• Not properly supervised by right/correct person or technical person • Repair work supervised by clerks • Aintenance and cleaning work not comprehensive supervised
(e) Lack of cleaning	12	• Involve for general cleaning work • Only surface cleaning executed at almost all schools • Cleaning executed not on technical based • Cleaning carried out by a not right person for example teacher or gardener

(continued)

Table 6.14 (continued)

Factors which have strong significance effect the building condition	No. of participants confirmed the significance	Reason(s)
(f) Misuse by user	12	• Misused of operation, for example door, sanitary fittings and fire by fighting equipment by either teacher, clerks, gardener, students or external user

...Without proper maintenance and technical person who carried out the school maintenance caused the condition become worst from day to day.

Meanwhile Participant E1 mentioned that less maintenance for example schedule maintenance lead to building deterioration. "...Schedule maintenance and standard procedure for maintenance is necessary to overcome building defects and building condition."

Secondly, for vandalism issue, 11 participants agreed that it has a strong significant relationship to the building condition. According to the interviews, almost all the vandalism issue happened to many schools but more to the school at rural areas. This issue become one of the school problems which involve destroying of the school properties. This also happened due to the student uncontrolled attitude that destroyed the properties and caused building defects. In the interview conversation, Participant E2 quoted that:

"....A lack of proper discipline regarding to the attitude toward vandalism will affect the physical condition of the school by producing damages and defects to the building elements such as damages to firefighting systems, sanitary fittings, door, window and many more." Meanwhile, only 1 participant neither agree the vandalism factor has significant relationship toward building condition due to cases and location of the schools.

Thirdly, for poor waterproofing factor, 11 participants agreed that poor waterproofing has strong significant relationship toward building condition. This involved the issue on the waterproofing warranty especially when involved leaking at the floor. Participant E1 quoted that "...Aging material waterproofing can cause leakage at floor especially toilet area." Besides that without proper waterproofing installed at school building can cause water to penetrate through wall and floor hence affect the condition of the building.

Meanwhile participant BS4 quoted that: "...Many of waterproofing problem is due to the expired of warranty."

Fourthly, for lack of supervision factor, 11 participants agreed the factor has a strong significant relationship toward building condition. Lack of supervision is either for maintenance, cleaning or repair work carried out by school management. Participant E2 quoted that:

> ...Many repair work or other contracts carried out by the school were not supervised by correct or right technical person. And this will produce shoddy workmanship to building element or fitting which can lead to the building defects.

Meanwhile, 1 participant disagreed that lack of supervision has strong significant relationship toward building condition. This is because, nowadays many school were designed to fulfil the requirement of the school organization without involving many alteration or upgrading work. The design completed with good quality material used for the school which reduce the issue of shoddy workmanship due to lack of supervision.

Then, all 12 participants agreed that cleaning factor has strong significant relationship which can affect the building condition. General cleaning work executed at almost every school is limited only to sweep the floor at teacher's room, and not comprehensive at other areas, for example toilet, which lead to building defects. The types of defects found at toilet area are blockage at floor trap, water ponding at floor, blockage at toilet bowl, and many more.

Participant BS5 quoted that "...Only surface cleaning was executed at school building, besides that cleaning are based on client requirement not on the technical based and most of it, cleaning is carried out without appropriate school procedure."

Lastly, all participants also agreed that misuse by user factor have strong significant relationship towards building condition. Without proper user or certain knowledge on the operation of the equipment by user can caused defects to the building. Participant BS2 mentioned that "...Misused of operation accordingly, for example door, sanitary fittings and firefighting equipment either by teacher, clerks, gardener, students or external user can caused building defects."

The interview result validated the ranking analysis result of strong significant factor that contribute to building defects towards building condition. Furthermore, the findings demonstrated the types of defects associated with the contributing factors that were not much covered in this research.

(ii) **Factors which have least significant and affect the overall condition of the building**

The ranking analysis for least significant factors is as shown in Table 6.15. Table 6.15 presents the interview results which supported the ranking analysis result of factors that contribute to building defects.

In this section, eleven (11) of twelve (12) participants were able to provide the information regarding the factors that contribute to the building defects which affect the overall building condition.

Participants confirmed that improper use of material factor has a least significant effect on the building condition. Wrong decisions for unsuitable material lead to building defects. However less case found for the wrong decision on the material selection.

This was supported by statement Participant E4, quoted that "...There is minor cases on material issues that contributed to defects after construction and affected the building condition."

Table 6.15 Investigation on least significant factors towards the building condition

Factors which have strong significance affect the building condition	No. of participants confirmed the significance	Reason(s)
(a) Improper use of material	11	• Minor issues on material affect the building condition • Wrong decision for unsuitable material lead to building defects
(b) Insect attack	12	• Almost happened at school which have wood material such as door and window frame, cupboard • Depend on location and treatment. No proper treatment will lead to the deterioration of the school buildings • Uncontrollable
(c) Faulty design	12	• Standard design for all school according to Malaysia Year Plan • Less for gap for improvement • Influenced factors lead to defects according to the location
(d) Wear and tear	12	• Wear and tear can be controlled to avoid defects to occur without repair work • No proper monitoring on the wear and tear factors can contribute to building defects
(e) Poor workmanship	12	• Issue on the workmanship chatting during construction/repair work • Since the school building are almost beyond repair, the workmanship factor least significance contribute to the building defects • Repair work carried out by not a right person/technical person
(f) Not complying with specification	11	• Wrong specification of material/system • Less issues on the specification

(continued)

Table 6.15 (continued)

Factors which have strong significance affect the building condition	No. of participants confirmed the significance	Reason(s)
(g) Reaction of thermal agent	12	• Changes in temperature which affect the material for example roof /wall of the school • Can cause thermal expansion and contraction which lead to the cracks
(h) Excessive moisture	12	• Uncontrollable, based on school operation and the usage of the building • Minor issues on moisture, happened not at all places at schools • Moisture can penetrate to building depend on its design
(i) Poor construction	12	• Improper construction lead to latent defects • Method of construction does not comply to the specification
(j) Change of usage	12	• Usage affecting the maintenance strength of school buildings • Misused of the schools building lead to building defects and affect the condition of the buildings
(k) Soil movement impact	11	• Many cracks at apron occurred at school due to soil impact due to least budget, ignore the soil improvement especially school
(l) Structural	12	• Structural issue due to aging building schools • Due to inter-related factor such as soil impact • Not many of school have structural problem/least significance

(continued)

Table 6.15 (continued)

Factors which have strong significance affect the building condition	No. of participants confirmed the significance	Reason(s)
(m) Change of climatic condition	12	• Climatic condition affect by location of schools • Heavy rain and temperature also effect the physical condition of schools which lead to the defects
(n) Reaction of chemical agent	12	• Influenced the properties of the materials when combine with extreme temperature • Lead to building defects when combine with moisture, oxygen, sulphate and another chemical agent
(o) Reaction of biological agent	12	• Fungus/fungi due to the dampness caused discolouration or dirty at building element especially wall • Bird droppings caused dirt and deterioration at school

Secondly, all participants agreed that insect attack by termite has a least significant relationship toward building condition. It almost happened at school which has wood material in construction such as door, window frames and furniture. The participants also agreed that the insect attack is one of the school problems and it depends on location and the frequency of treatment done by schools. In the interview conversation, Participant BS1 quoted that:

> … No proper treatment will lead to termite spread uncontrollable and caused the deteriorations of the school building.

Thirdly, faulty design factor has a least significant relationship which affects the building condition. Not many issues reported on the school building has defects due to the design. All participant agreed that design factor has least significant relationship towards building condition. This statement was supported and explained during the interview session.

Participant E2 explained that "… Not many defects reported for school caused by design. This is because, school has a standard design as planned in Malaysia Year Plan so that the gap for improvement is less."

> … Design factor lead to the defects according to the location, but it is a minor case.
> (Participant BS5)

Fourthly, wear and tear factor contributed to building defects and affected the overall building condition. Wear and tear can be controlled to avoid defects to occur without repair work. For school building, wear and tear factor has least significant relationship toward building condition. School was designed with the use of materials that can resist abrasion with high wear and tear. All participants agreed that this factor has least significant toward building condition.

> The interviewee explained that "…No proper monitoring and awareness on the wear and tear issue especially for school user can contribute to building defects."
>
> (Participant E6)
>
> … Defects were due to occupant or the user of the building and most related to wear and tear such as damaged or rusty ironmongery, sagging or in alignment of material such as door.
>
> (Participant BS2)

Then, poor workmanship factor has a least significant relationship with building condition. The issue on the workmanship chatted only during construction or during repair work. All participants agreed that workmanship has a significant relationship with overall condition of the building.

> The interviewee explained that "…Many schools in Malaysia have their repair work after 10th year old, so many issues on the workmanship are found especially during renovation or repair work. This happened when to suit the new work with existing element/structure of the building."
>
> (Participant BS2)
>
> … Since the school building are almost beyond repair, the workmanship factor does not become a main factor to contribute to building defects.
>
> (Participant E1)
>
> … Repair work carried out by not a right person or right technical person especially for school buildings. That is why workmanship factors in gives effects to condition of the buildings.
>
> (Participant BS5)

For not complying with the specification factor also has least strong significant relationship which affect the building condition. Eleven (11) of participants agreed that specification has least significant relationship towards building condition. Meanwhile only one (1) participant disagreed with the statement.

> … There are minor cases of defects because of specification. This is due to the suitability of the new system to suit with the existing building (after occupancy) which is more than 10 years old. The right specification of the material or system chosen is very important to prevent defects to occur.
>
> (Participant E5)
>
> … Almost at school, not many defects happened due to the specification, but it is due to lack of supervision, skill during construction or repair work. Many school in Malaysia have standard design and specification.
>
> (Participant BS1)

Changes in temperature due to reaction of thermal agent affect the building material especially for school buildings. This happened to the school building with the age of more than 10 years old. Thermal agent also can cause thermal expansion and contraction which lead to building cracks especially to wall. Most of the participants agreed that thermal agent has a least strong relationship that affected the overall building condition. Meanwhile only one (1) participant neither agreed that thermal agent has a least significant relationship toward building condition.

> In the interview, the participant explained that "…Thermal agent caused changes in temperature lead to vertical cracking in walls and also problem with roofs, cladding and external structure members."
>
> (Participant BS3)
>
> … Thermal agent only gave effects on jointing material and it found less at school buildings.
>
> (Participant E3)

All participants agreed that excessive moisture is uncontrollable when involved to climatic condition. But moisture can be prevented based on design and material used during construction.

For school building with the age more than 10 years old, majority of the building have moisture problem due to environment, school operation and the usage of the building. Moisture comes whether from rain, condensation or water penetrated through wall into the building.

Participant E6 explained that "…Moisture at school is a minor issue and it depends on the design of the building. Almost school building follows a standard design and the circumstances to moisture trapped in the building are less especially for new design of school building."

> … Moisture in building happened at many schools, and it depends on the location of the school, and age of schools.
>
> (Participant BS4)

Almost all the interviewees have a similar prospective and explanation on the poor construction factor. For poor construction factors, all participants agreed the factor that has least significant relationship toward building condition. Without proper supervision, latent defects occurred during construction.

Participant B3 explained that "…Defects during construction can be seen after occupancy and it is due to the method of construction, without proper supervision and also not compliance to the specification."

> … Defects due to the construction factor is not a major contributor affecting the building condition because of their standard design especially for school.
>
> (Participant E4)

Change of usage of school factor also contributed to the building defects and has a least significant relationship toward building condition assessment. In the interview, Participant BS1 explained that: "…The wrong usage of school whether for school space, building element, systems or components, most of it affecting the maintenance of the school." Meanwhile, Participant E5 explained that wrong usage of the schools building lead to building defects and affect the condition of the buildings for example misused of fittings, doors, and other school furniture desk and chairs.

Soil impact movement factor also has least significant relationship to building condition due to minor cases found in the study where eleven (11) participants agreed to the statement as found in the study. The movement of soil can give an impact to the building element such as crack at apron besides shoddy workmanship. Participant E1, mentioned that: "…Many cracks at apron occurs at school due to less budget on soil improvement." In the interview also, Participant BS6 explained that: "…Crack at wall, and apron is due to soil movement. This happened when cut and fill soil work are not compacted according to the requirement before the construction of the school. But this is only minor cases."

Structural factor sometimes can be a big issue to the school problems. During the interview session, the participants explained that the structural issue is due to inter related factor such as soil problems. Structural issue happened to school with age of more than 10 years. Eleven (11) participants agreed that the structural factor has a least significant relationship toward building condition. Meanwhile only one (1) participant neither agreed based on the explanation:

> … Not many schools have the structural problem. Standard design for school in Malaysia is less than 5 stories in heights, so that the structural issue can be controlled.
> (Participant E4)

> … School with age of more than 10 years old has potential to structural issue, anyhow it depends on location and force majeure.
> (Participant BS3)

All participants agreed that change of climatic condition has least significant effects on the condition of the building. Climatic condition affects by location of schools. The heavy rain and temperature within school area affect the physical condition of schools. Many defects can be found due to climatic condition factor for example algae, fungus, dampness and many more. According to participant BS5, climatic condition due to heavy rain caused defects problem at almost all schools especially dampness to wall and floor.

> … Water or rains penetrate to the building without any warning or sign at early stage and it appears after a certain time. This happened at school more than 10 years old.

Reaction of Chemical agent factor has least significant relationship toward building condition. All twelve (12) participants agreed that this factor has least significant relationship because the chemical agent influenced and gave affect to the building element, structure or system when it combined with extreme temperature. Participant BS3, explained that: "…Chemical agent will lead to building defects when combined with moisture, oxygen, sulphate and other chemical agent."

For reaction of biological agent factor, all 12 participants agreed that biological agent has least significant relationship toward building condition. Biological agent for example fungus/algae caused discoloration or dirty at building element especially wall. Other than that, bird droppings also caused dirt and deterioration at school. In the interview conversation, participant E1 quoted that:

> … Almost school which with aged more than 10 years have this lot of problem and become uncontrollable if left unattended for a long time.

The interview results validated the ranking analysis of factor which has least significant relationship toward building condition assessment. None of the participant suggested any new factors that are not covered in this research.

(iii) Factors which have zero variance to the overall condition of the building

In this section, twelve (12) participants were able to provide the information regarding to factor that contribute to the building defects which affect the overall building condition. The investigation analysis for factor which have zero variance towards building condition is as shown in Table 6.16.

There are eight (8) factors ranked as zero variance in this study. The factors have been removed from the first analysis since those factors are not much related to the study. In the semi-structured interview, all participants confirmed the factors which have zero variance toward building condition. Detailing as a factor which contributed to building defects are not related for existing buildings especially after occupancy of the building.

> The interviewee explained that "… Poor detailing factor is not applicable for school building since the design of school is not complicated and a standard design for all school in Malaysia. Detailing factor also does not affect the condition of the building."
>
> (Participant E1)
>
> … Not many schools have complicated detailing especially for the structure or building element which can lead to the building defects.
>
> (Participant BS2)

Secondly, all participants agreed that drawing has zero variance which can contribute to defects to the school building. The availability of the drawing is not appropriate since the school in Malaysia all used standard design. The drawing is not a factor which can contribute to the building defect for existing building. Hence the drawing also is not kept by the school management.

In the interview, Participant BS4 explained that:

> … Building defects cannot trace the defects due to drawing fault especially for building of more than 10 years old.

Thirdly, all participants also agreed that building size has zero variance toward building condition. The defects can be found without considering the size of the building. Most of the school building in Malaysia are inspected by PWD and have different types of defects which affect the building condition due to other factors. The building size can contribute to the volume of the defects. Participant BS3 explained that:

> … Most government building including schools faced defects problem without considering the size of the building either single storey building or multi storey building.

All participants agreed that the building type factor has zero variance towards building condition. For building type factors, schools have the same standard and types of design. In this study, the school was selected which have same design according to the criteria set up for the study. Thus, the type of defects can be projected to be the same for each school according to the location. In the interview session, participant E6 explained that:

Table 6.16 Factors which have zero variance toward building condition

Factors which have zero variance affect the building condition	No. of participants confirmed the significance	Reason(s)
(a) Poor detailing	12	• Not appropriate/not affect the condition of the building • Not many schools share complicated detailing to the building element which lead to building defects
(b) Faulty drawing	12	• Not appropriate • Not kept by school/cannot be traced the defects is due to drawing fault especially for building more than 10 years
(c) Building size	12	• Building size does not matter only for one (1), two (2), three (3) storey height of school building • Almost all of the building inspected by PWD have different types of defects which affect the building condition
(d) Building types	12	• Different occupant/user • Same operation • School has same types of design/standard building
(e) Building orientation	12	• Building orientation requirement included in the brief
(f) Lack of knowledge	12	• Students/pupils/teachers is non—technical person which have no knowledge on building defects
(g) Poor ventilation	12	• Not appropriate at school building, all schools have natural opening for natural ventilation hence avoid defects • Schools completed with mechanical ventilation such as fan • Defects to element does not affect the students/pupils

(continued)

Table 6.16 (continued)

Factors which have zero variance affect the building condition	No. of participants confirmed the significance	Reason(s)
(h) Poor operation	12	• Operation is depended on the standard procedure especially when operating the equipment. In this case school did not have many equipment/machineries during the operation of the school

… Each school has different occupant or user with same operation. So, the defects are not caused by the types of building, but it is depending on the activity of the user.

Through the interview session, all participants confirmed that building elevation is one of the factors which has zero variance significant relationship toward building condition. Building elevation requirement included in design brief at early stage of planning project by considering the wind direction, light and others.

The building elevation do not affect the defects in the building. In the interview session, Participant BS4 explained that:

… The defects found at school were not influenced by building elevation or type of building, but the environment factors would lead to building defect.

Lack of knowledge is one of the factors that can contribute to the building defects. But for the case of school building, all students/pupils/teachers were non –technical person which have no knowledge on the building defects. All participants confirmed that this factor has zero variance towards building condition. This is due to most of school which do not have technical person who will take care of the building. Thus, the school design has to suit the user of the building. Participant E6 explained that:

… Teachers or students have zero knowledge about the defects and factors that contribute to the building defects. So that the lack of knowledge is not applicable for user, but it is more to the contractors or supervisors who executed the renovation, upgrading work.

Without proper ventilation can caused defect to the building. In this study it is found that ventilation factor has zero variance or no significant relationship towards building condition. As conclusion from interview session, all participants confirmed that the ventilation factor is one of the factors that has zero variance which can affect the building condition.

In the interview sessions, Participant BS5 explained that:

… Ventilation factor is not a reason for school building decay. This factor is not appropriate for school building. All schools have natural opening for natural ventilation to avoid defect.

According to Participant B5:

> … School buildings are completed with mechanical ventilation such as fan to ensure the proper flow of air within space to give comfort from heat or other factors which do contribute to building defects.

Operation of the school depended on the standard procedure especially when operating the equipment or machinery. For school building the operation of the school remains the same for all schools without any operation on heavy equipment or machineries. All the participants confirmed through the interview session that operation factors were not the reason the school condition deteriorate.

Participant BS6 quoted that "… Most of the schools have one session of school per day. So the operation and the uses of school is less than two session of school per day. The operation of school is not the main reason as factor that contribute to the building defects."

Therefore, the interview results validated the ranking analysis of factors which have zero variance which can affect the building condition. None of the participant suggested any new factors that are not covered in this research.

(iv) **Theoretical framework**

The correlation analysis identified the most significant factors that contributed to the building defects. To isolate and identify the most likely causes based on the study, the theoretical framework of the actual problem as possible independent, mediating and dependent variables was developed.

(a) **Independent Variables**

The interviewer explained the theoretical framework of the study and asked about the flow and significant relationship between factors and building condition. Table 6.17 shows that the reason and suggestion by participants to confirm the variables in the theoretical framework.

In this study, the factors categorized under three (3) main factors which are technical, environmental and human factor. The factor was grouped based on previous research and to be confirmed and validated by the participants. All the participants confirmed that the independent variable identified would lead to building defects.

Participant E1 explained that:

> … The technical, human and environmental factors can contribute to the building defects and affect the condition of the building. Most of the government buildings are influenced by these factors as reported by PWD.

Participant E4 explained that "…The categorization of factors according to the similarity to ease the updating of current government's inspection system. So, it is better if the factors were grouped into the similarity based on the research conducted."

The interviewers also ask the relationship between factors that can contribute to building defect hence affect the building condition. All participant confirmed and agreed the combination of factors can lead to different extent of defects. In this case one participant give a few examples of defects according to the combination of factors.

Table 6.17 Reasons and suggestions to the variables

No	Variables	No. of participant (confirmed the variables)	Reason(s)/suggestion(s)
1	Independent variables • Factors contribute to building defects	12	• The factors that contribute to building defects will affect the building condition • The various factors related to each other to contribute to various type of building defects • Bank of factors is important since not all factors causes the building defect • The list of factor are commonly causes building defects in Malaysia • The list of factors to be updated in the government's inspection system to enhance the existing processes • The categorization of factor according to the similarity to ease the updating of current government's inspection system • The factors contribute to building defects have significant relationship affect the building condition
2	Mediating variables • Type of defects	12	• Type of defects effects the overall condition of the buildings according to its degree • Types of defects will be different and depends on the type, location and uses of the buildings • The type of the defects are common defects and are usually found for government building in Malaysia • The list of defects to be updated in the government inspection system to enhance the existing processes • The relationship between defects and building condition should be studied for further improvement

(continued)

Table 6.17 (continued)

No	Variables	No. of participant (confirmed the variables)	Reason(s)/suggestion(s)
3	Dependent variables • Building condition	12	• Building condition reflects the value of the building • Building condition for monitoring and maintenance prediction projection cost for future • For planning maintenance program • Improve the maintenance processes • Client satisfaction

> … The various factors related each other to contribute to the various types of defects. For example, the user factor can be combined with usage factor results wear and tear and also lead to defect such as damages, peel off and others.
>
> (Participant BS2)

Another participant explained that:

> … The framework is easy to interpret the various factors in the list that can contribute to the different types of defects such as the material factor combination with climatic condition factor could lead to decay, mold, fungi peel off to the building element.
>
> (Participant BS5)

The interviewer asked about the importance of all factors to be known during the inspection to ensure the proposal of repair work based on the source of the defects. With these factors, the right remedy action would be taken within a reasonable cost of repair or maintenance.

In this case Participant BS3 quoted that:

> … The bank of factors should be developed or updated in the current guideline to be used during inspection to all types of building. Not all factors cause the building defect but it depends on location and also the type of building.

The factors identified in the research are important to be a reference for all types of buildings. In this case, the factors are only used for the school building. Although not all factors are found at school buildings, but the list of factors should be updated in the government system to enhance the existing process of inspection and maintenance.

The results validated the ranking of factors which have significant relationship toward building condition. Furthermore, the findings demonstrated the types of defects associated with the contribute factors that were not much covered in this research.

(b) **Mediating Variables**

There are various types of defects found in the research. Hence those types of defects are different according to building location, the use and the extent of the defects. Defects can be caused by the combination of the factors as a source of defects. All the participants agreed that the mediating variables indicated in this study are common defects found in buildings. One participant explained that:

> … The types of defects found in the study are common defects for government building in Malaysia. Those type of defects should be listed and updated for future references.
>
> (Participant E2)

Another participant suggested that:

> … Most of the defects stated in the study are found and reported in PWD reports. The list of defects should be updated in the government inspection system to enhance the existing process.
>
> (Participant BS6)

Thus, the interview result validated the mediating variables in the theoretical framework. Most of the participants suggested to update the list of defects for future references. The building rating dictionary should be developed to ease the process of inspection and maintenance; since this study only focused on the relationship defect factors and building condition, the relationship between defects and building condition should be studied for further improvement.

(iii) **Dependent Variables**

The information of building condition is to forecast possible future repair or maintenance requirement. That is why the condition of the building is very crucial and important to ensure the beneficial aspect of the building. All the participants agreed and confirmed that dependent variable, building condition has a significant relationship with the defects factor. In the interview one participant quoted that:

> … The building condition is important to ensure safety, satisfaction and comfort to the user or client. Hence it also reflects the value of the building.
>
> (Participant E6)

Another participant pointed out that:

> … Building condition for monitoring and maintenance prediction projection cost for future. So that to maintain the building condition is important and crucial to maintenance planning which are related to cost of repair.
>
> (Participant BS3)
>
> … The importance of building condition is planning for maintenance program to improve the maintenance processes.
>
> (Participant BS2)

The interview result validated the dependent variables of the study. Furthermore, the importance of the condition assessment provides an opportunity for future study on the defect that affect the condition of the building.

(xxii) **The Uses of Standard Building Condition Assessment in the Study**

To assess the condition of the building, the guideline and system of government BCA were adopted in the study. BCA rating are commonly used as weighing factor to represent the severity of the defects. The use of BCA needs to be validated to ensure the practicality and workability of BCA. During the interview session, all the participants confirmed that the use of the BCA is comprehensively practical especially for all kinds of buildings.

The interviewer asked the participants about their opinion on the practicality of the uses of BCA guideline. One participant explained that:

> ... BCA is practical and good as inspection tool for diagnosis of the building. The overall ratings of the building reflect exact condition, maintenance and repair costs for the building.
>
> (Participant E5)

Another participant noted that:

> ... BCA is a step by step approach to get the building rating. So that it helps to ease the inspection processes. Beside that BCA is also as a tool in inspection for standard building to come out with a lesson-learnt for future design, further action and others.
>
> (Participant E6)

Another participant agreed and stated that:

> ... BCA ease the inspection besides to ease the preparation of repair costs including the rectification work. BCA also provide the platform of prediction for future condition, cost and maintenance planning.
>
> (Participant BS1)

The interviewer also requested the suggestions from the participants to enhance the current processes. Majority of them agreed that it is necessary to enhance the building condition rating according to the factors and types of defects. During the interview session, participant BS2 explained that:

> ... BCA is a comprehensive tool but it needs to be improved and reviewed for inconsistency in the finding and priority maintenance assessed by inspectors.

The interviewer also has to confirm the method on how to rate the extent of the defects. The participants confirmed that the method to carry out the inspection based on the BCA guideline. Participant E5 explained that:

> ... To rate the extent of the defects is through observation of physical condition and priority maintenance assessment. Anyhow it depends on the size of the defects which reflect to proposal of rectifying work and cost.

To ensure and to confirm the accuracy of the building rating, another participant explained that:

> ...The accuracy of the finding during the inspection are based on proper training, experiences and knowledge.
>
> (Participant BS6)

Another participant suggested:

> ... Follow the guideline of BCA step by step, and other statutory enforcement such as Uniform Building by Law and also forensic guideline.
>
> (Participant E4)

Table 6.18 The use of BCA

No	Description	Reason(s)/suggestion(s)
1	The opinion on the practicality of the uses of BCA guideline	• General diagnosis of the building • Ease the inspection processes • Good for references and for rectification work • Step by step approach to get building rating • Proposed procedure to ease the inspection • As a tool in inspection for standard building to come out with a lesson learnt for future design, further action and others • Ease to prepare the repair cost Prediction for future cost of maintenance planning and cost
1		• Need to be improved and reviewed for inconsistency the finding and priority maintenance assessed by inspectors • Need to enhance the building condition rating according to the factors and types of defects
2	How to rate the extent of the defects in the inspection	• Through observation to the level of physical condition and priority maintenance assessment • Assessment on physical condition of the building and priority maintenance based on the finding • Depend on the size of the defects • Based on matrix analysis for physical condition and priority maintenance action • Procedure as in the guideline
3	How to get the accuracy of the building condition rating	Referring to: • BCA guidelines • Building rating dictionary • Experiences • Knowledge • Expert judgement • other statutory—uniform building by law and others • Drawing • Forensic guideline • Proper training

Table 6.18 shows the points highlighted by the participants regarding the use of BCA. The participant pointed out some reasons and suggestions according to the interview questions.

The result of interview confirmed that the use of BCA tool in this study is practical to improve the current inspection and maintenance processes especially for government building. The result also confirmed the method to rate the extent of the building defects used in this study. The results also confirmed the accuracy of the building condition rating in this study.

6.7 Discussion on Overall Findings

The triangulation method used in this study to validate and confirm the quantitative results with qualitative findings. The data were collected, analysed, merged, compared results and interpreted the data and findings. Two methods were used to check the results of the findings. This includes condition assessment through archived documentation and semi-structured interview. The main research findings were gained from condition assessment, and then the findings would be compared with the interview results for validation. The validation covers the results of ranking analysis, correlation analysis and theoretical framework of the study.

6.7.1 Validation of Ranking Analysis Results

The ranking analysis result for the factors that contribute to building defects revealed three (3) of the most significant factors that contributes to the building defects. The highest rank factor is lack of supervision. The second highest rank is lack of maintenance, followed by vandalism factor. The interview results demonstrated that all participants confirmed these three (3) significant factors. The analysis also confirmed that the factors have a significant characteristic to contribute to building defects.

6.7.2 Validation of Correlation Analysis Results

The correlation analysis result revealed that there are six (6) factors which have strong significance affect to the condition of the building. These include lack of maintenance, vandalism, poor waterproofing, lack of supervision, lack of cleaning and misuse by user. In the interview session, all participants confirmed that the lack of maintenance, lack of cleaning and misuse by user factor have a strong significant relationship towards building condition. Then, 92% of the participants revealed the vandalism and lack of supervision have a strong significant relationship towards building condition. Meanwhile 83% of the participants agreed poor waterproofing has a strong significant relationship towards building condition.

Furthermore, the correlation analysis results for factors have least significance affect to building condition demonstrated fifteen (15) that are significantly correlated to building condition. All the interview participants acknowledged insect attack, faulty design, wear and tear, poor workmanship, reaction of thermal agent, excessive moisture, poor construction, changes of use, structural, changes of climatic condition, reaction of chemical agent and reaction to biological agent have least significant relationship to building condition. In addition, 92% of participants confirmed that improper use of material, soil movement and not complying with specification significantly have least correlation towards building condition.

For factors which have zero variance toward building condition, the correlation analysis result demonstrated that eight (8) have zero variance toward building condition. These factors are poor detailing, faulty drawing, and building size, building type, building orientation, lack of knowledge, poor ventilation and poor operation. The factors have been removed from the first analysis since these factors are not much related to the study. All interview participants confirmed the factors are not related to the case of building school.

In general, the interview and the result findings through condition assessment and archived documentation result validated the correlation analysis.

6.7.3 Validation of Theoretical Framework

Literature shows that the factors contributing towards building defects are related to the building condition. Evidence of outcomes are mediated by types of defects found in the study. For validation of theoretical framework of the study, factors which are as independent variables were grouped into three (3) main factors; technical, environmental and human factor. Mediating variables were types of defects found in the research, twenty eight (28) types of building defects and ten (10) types of mechanical and electrical defects were found in the study.

In this study, the description of [], construct validation framework and corresponding steps used to develop and validate the theoretical framework is shown in Fig. 6.7. First, theoretical and empirical study are researched and developed. From the literature, it is found that 40 factors caused building defects. The items were selected as variables and arranged to similarity and grouped into 3 categories which is technical, environmental and human factors consisting of 29 factors used in the study. The items were then tested for their reliability and consistency to check their relationship towards the building condition. Secondly, the relationships between the factors contributing towards the building defects towards building conditions were examined using correlation analysis.

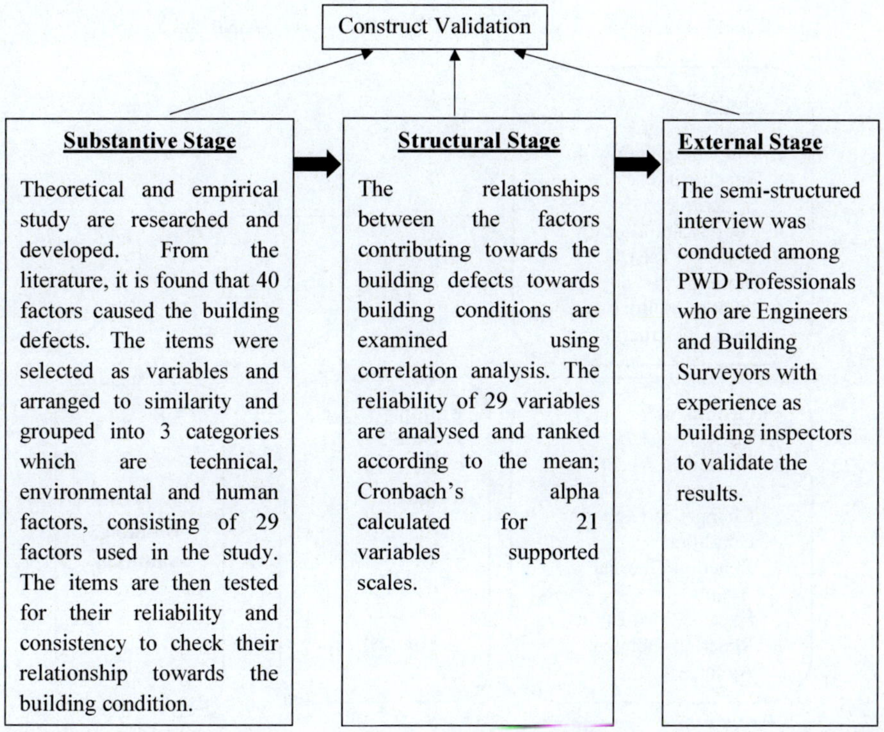

Fig. 6.7 Validation framework

The reliability of 29 variables were analysed and ranked according to the mean; Cronbach's alpha calculated for 21 variables supported scales. For the last stage, the semi-structured survey was conducted among PWD Professionals; Engineers and Building Surveyors with experience as building inspector to validate the results. Figure 6.8 shows the final theoretical framework produced in the study. The building defects are influenced by the 3 categories of factors contributing towards the building defects.

In general, the findings validated the variables and framework of the study. All interview participants comfirmed and agreed to the arrangement of the factors into categories according to the similarity, and confirmed that those defects were various types of defects found at school buildings.

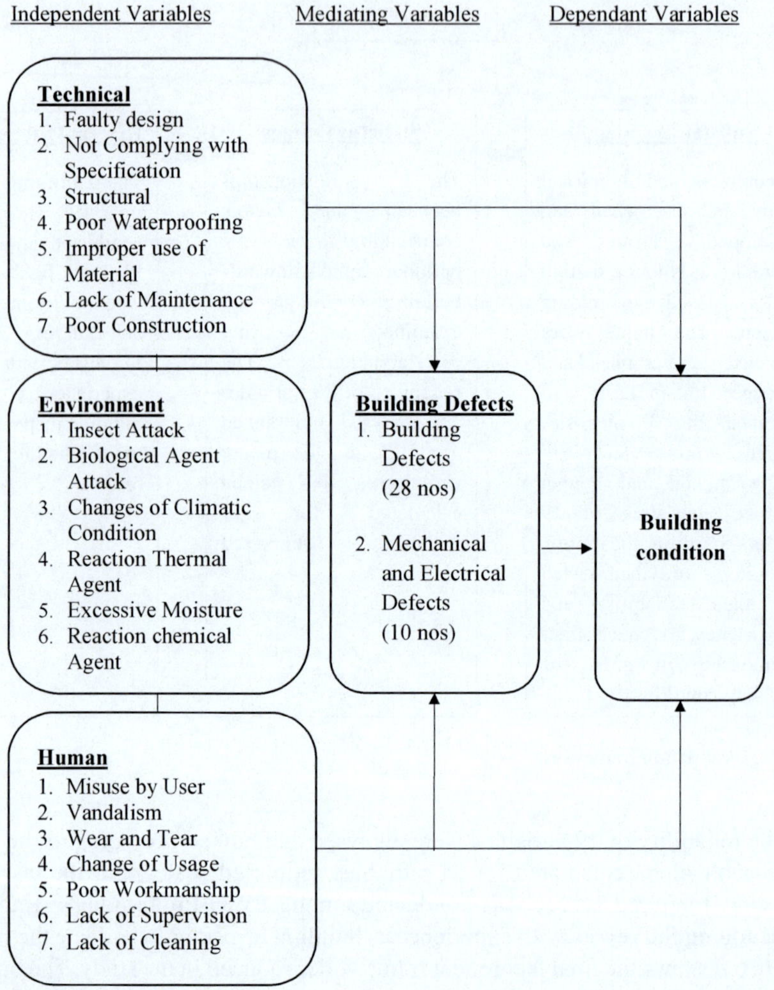

Fig. 6.8 Final framework

References

1. PWD. (2013). *Building condition inspection for existing building*. Malaysia Public Work Department.
2. Zamri, M. A. H., & Abdullah, J. (2018). Evaluation of infill building block in heritage schools in Kuala Lumpur. *Planning Malaysia, 16*(7).
3. Roslan, T., Ahmad, A. G., Zakaria, N., & Sulieman, M. Z. (2014). Assessment of factors affecting building maintenance and defects in public building in Penang Malaysia. *Architecture Research*, 48–53.https://doi.org/10.5923/j.arch.20140402.03
4. Macabuag, J., Bhattacharya, S., & Blakeborough, T. (2008, October). Extending the collapse time of non-engineered masonry buildings under seismic loading. In *Proceedings of the 14th World Conference on Earthquake Engineering* (pp. 12–17).

5. Noy, E. A. (2005). *Building surveys and reports.* B00Z8GJ3WW, Wiley-Blackwell (January 1, 1700).
6. Che Ani, A. I., Sairi, A., Johar, S., Razak, M. Z., & Tawil, N. (2015). The practice of building inspection in Malaysia total asset management. *Jurnal Teknologi*, 1–11.
7. Choudhry, R. M., Gabriel, H. F., Khan, M. K., & Azhar, S. (2017). Cause of discrepancies between design & construction in the Pakistan construction industry. *Journal of Construction in Developing Countries, 22*(2), 1–18.
8. Kudder, R. J., & Erdly, J. L. (Eds.). (1998). *Water leakage through building facades.* ASTM STP 1314, West Conshohocken, PA 19428-2959.
9. National Research Council. (1985). Building diagnostics: A conceptual framework. The National Academies Press. https://doi.org/10.17226/19294
10. Fire Services Regulation, The Institution of Fire Engineers (UK) Malayasia Branch (IFEM). ISBN 983-808-095-0.
11. Mohsen, T. (2011). Making sense of Cronbach's Alpha. *International Journal of Medical Education*, 53–55.
12. Waziri, B. S. (2016). Design and construction defects influencing residential building maintenance in Nigeria. *Jordan Journal of Civil Engineering*, 313–323.
13. Assaf, S., Al Hammad, A. M., & Al-Shihah, M. (1996). Effect of faulty desin and construction on building maintenance. *Journal of Performance Construction and Facilities*, 171–174.
14. Okuntade, T. F. (2014). Effects of faulty design and construction on building maintenance. *International Journal of Technology Enhancements and Emerging Research*, 59–64.
15. Agwu, M. O. (2014). Perception survey of poor construction supervision and building failures in six major cities in Nigeria. *British Journal of Education, Society & Behavioural Science*, 456–472.
16. Ahzahar, N., Karim, N., Hassan, S., & Eman, J. (2011). A study of contribution factors to building failures and defects in construction industry. In *Procedia Engineering. The 2nd International Building Control Conference* (Vol. 20, pp. 249–255). Elsevier. https://doi.org/10.1016/j.proeng.2011.11.162
17. Ali, A.S., Keong K.C., Zakaria, N., Zolkafli, U., Akashah, F.(2013), The Effect of Design on Maintenance for School Buildings in Penang, Malaysia," *Journal Structural Survey*, 194–201.
18. Suffian, A. (2013). Some Common Maintenance Problems and Building Defects: Our Experiences, *Elsevier Procedia Engineering Science Direct*, 101–108
19. Wardhana, K. (2003). Study of Recent Building Failures in the United States. *Journal of Performance of Constructed Facilities,* 151–158.
20. Chew, M. (2004). Maintainability of Wet Area of Non Residential Building. *Journal Structural Survey*, 39–52.
21. Constantinos, A. B. (2005). *Deterioration of European Apartment Buildings.* Energy and Buildings, 515–527.
22. Md Ulang, N. (2014). *The Impact of Materials and Maintenance Considerations During the Design Stage of Public Buildings in Oman.* SHS Web of Conferences, 1–6. EDP Sciences.
23. Othuman Mydin, M. A., Ramli, M., & Awang, H. (2012). Factors of Deterioration in Building and the Principles of Repair. *Analele Universitatii, ISSN, 1453–7397*, 345–352.
24. Talib, R. (2014). Assessment of Factors Affecting Building Maintenance and Defects of Public Buildings in Penang, Malaysia. *Journal Scientific Academic Publishing*, 48–53
25. Adenuga, O. A. (2012). *Professionals in the Built Environment and the Incidence of Building Collapse in Nigeria.* https://doi.org/10.5592/0tmcj.2012.2.2ResearchPaper,461-473
26. Bagdiya, N. V., & Wadalkar, S. (2015). Review paper on construction defects. *IOSR Journal of Mechanical and Civil Engineering (IOSR-JMCE)*, 88–91
27. Meyer, H. H., & Raich, M. S. (1983). An objective evaluation of a behavior modeling training program. *Personnel Psychology, 36*(4), 755–761.
28. De Wet, C. (2004). The extent and causes of learner vandalism at schools. 24. https://doi.org/10.4314/saje.v24i3.24990.

29. Johnson, Kelly Dedel. 2005. School Vandalism and Break-Ins. Problem-Oriented Guides for Police Problem Specifi c Guides Series, No. 35. Washington, DC: U.S. Department of Justice, Offi ce of Community Oriented Policing Services. Available: www.popcenter.org/Problems/problem-vandalism.htm.
30. Jaafar, M., Othman, N.L. (2015). Categorisation and Causes of Building Design Defect: A Case Study on Public Building Hospital.
31. Yacob, S. (2005). *Maintenance Management System Through Strategic Planning for Public School in Malaysia*. UTM.
32. Chong, W.K., & Low, S.P. (2006), "Assessment of Defects at Construction and Occupancy Stages," *Journal of Performance of Constructed Facilities*, 213–221.
33. Morton, J. (2012, 6 21). *3 ways to Protect against Vandalism*. Retrieved from Buildings Smarter Facility Management.
34. Daisey, Joan M., William J. Angell, Michael G. Apte, 2003. Indoor air quality, ventilation and health symptoms in schools: an analysis of existing information. Indoor Air 13 (1), 53–64.)
35. Sreenivasan, J., Govindan, M., Chinnasami, M., & Kadiresu, I. (October 26th 2012). Solid waste management in Malaysia—a move towards sustainability, waste management—An integrated vision. Luis Fernando Marmolejo Rebellon, IntechOpen. https://doi.org/10.5772/50870. Available from: https://www.intechopen.com/books/waste-management-an-integrated-vision/solid-waste-management-in-malaysia-a-move-towards-sustainability
36. Annila, P. J., Hellemaa, M., Pakkala, T. A., Lahdensivu, J., Suonketo, J., & Pentti, M. (2017). Extent of moisture and mould damage in structures of public buildings. *Case Studies in Construction Materials*, 103–108.
37. Khan, M. M. H. (2019). Effect of temperature and relative humidity on the population dynamics of brinjal and tomato infesting whitefly, Bemisia tabaci. *Jahangirnagar University Journal of Biological Sciences., 8*, 83–86. https://doi.org/10.3329/jujbs.v8i1.42471
38. Hong, C. H. (2016). *Investigation of defects in new buildings in Malaysia*. Universiti Tunku Abdul Rahman.
39. Abas, A. A. (2014, 8.12). Building maintenance. Skudai, Johor, Malaysia: Engineering.
40. Hamzah, N., Mahli, M., Che Ani, A. I., Tahir, M., Abdullah, N., & Tawil, N. (2010). The development of smart school condition assessment based on condition survey protocol (CSP) 1 matrix: A literature review. *Science, Engineering and Technology*, 620–625.
41. Allison, P. (1990). Change scores as dependent variables in regression analysis sociological methodology. 20.https://doi.org/10.2307/271083
42. Benson, J. (2005). Developing a strong program of construct validation: A test anxiety example. *Educational Measurement Issues and Practice, 17*(1), 10–17.

Chapter 7
Conclusion and Recommendation

Abstract This chapter presents the conclusion of the study and followed by recommendations for future research. This research associated by reviewing the literature to identify and formulate the objectives of the study. The conclusion of the research drawn from the results of condition assessment, semi-structured interviews and archived documentation then form the basis of recommendation for further research.

7.1 Conclusion of the Study

Many school buildings in Malaysia are aging, and sustaining their condition and operation has become a great challenge particularly in maintaining the building which complicated decision making on the maintenance budgets. The decision are highly dependent on accurate condition assessment. An in-depth understanding about the building condition and the factors that influence the defects were studied through literature review. The factors were determined in the study to analyse the significant relationship toward building condition. Then, the condition of the building were analysed accordingly to check the reliability of the condition assessment results. A theoretical framework developed to contemplate the structure of the study.

A triangulation approach was used in the study involving data collected through condition assessment and content analysis from archived documentation. The data were then analysed together with the semi-structured interview results to compare and validated both results. In the conclusion, the factors that contribute to building defects are significant toward the building condition. The conclusion from the study are discussed in the following subsection.

S. Yacob et al., *Managing Building Deterioration*, Management in the Built Environment,
https://doi.org/10.1007/978-981-16-5860-0_7

7.1.1 *Objective 1: To Identify the Factors Contributing Towards Building Defects*

The factors that contributed to building defects were identified through literature review and archived documentation. There are 40 factors found in the study. These factors are grouped into similarity and a total of 29 factors were adopted in the study. Only 21 factors were ranked for most factors that contribute to building defects. The factors identified have been analysed using relative index (RII) to determine the relative importance of the factors causing the building defects. The factors that contribute to building defects for the study are stated in Table 7.1. The factors were sorted in descending significance rank.

As a conclusion, the factors which contributed to building defects are different and is based on the location and types of building. A similar study on the different types of building is necessary to ensure all factors listed in the study has a significant relationship with building condition.

Table 7.1 The list of factors that contribute to building defects

No	Factors
1	Lack of supervision
2	Lack of maintenance
3	Vandalism
4	Improper use of material
5	Poor workmanship
6	Wear and tear
7	Poor construction
8	Excessive moisture
9	Misuse by user
10	Poor waterproofing
11	Change of climatic condition
12	Reaction of biological agent
13	Insect attack
14	Soil movement impact
15	Lack of cleaning
16	Reaction of chemical agent
17	Faulty design
18	Change of usage
19	Reaction of thermal agent
20	Structural
21	Not complying with specification

Table 7.2 Percentages of overall condition rating

Rating	A	B	C	D	E
Percentage (%)	0	5.33	72.67	21.67	0.33

7.1.2 Objective 2: To Investigate the Condition of Public School Buildings Through Building Condition Assessment

The condition assessment were carried out and archived documentation were studied and analysed to identify the condition of the building. The condition assessment carried out based on the PWD BCA guideline with a first step referring to the inspection. Then the next process is identifying the defects or failure, followed by assessment to physical condition and maintenance priority. The grade of the assessment then were transferred to defect score (BCARS) to get the final building condition rating. The overall condition of the building according to the rating is shown in Table 7.2.

7.1.3 Objective 3: To Establish Critical Factors Contributing Towards School Buildings Defects and Its' Relationship to Building Condition

The significant relationship factors towards building condition were identified through condition assessment result, semi-structured interview and archived documentation. The correlation analysis was used to analyse and to check the significance relationship between independent variables and dependent variables.

A total of twenty one (21) variables was tested with condition of a building. According to the result of analysis, it is proven that the factors that contribute to building defects are significantly correlated to the condition of the building. The significant variables as shown in Table 7.3.

In conclusion, the factors which have a significant relationship influenced the condition of the building. This factor should be taken into consideration in order to minimize or eliminate the defect of the building.

Table 7.3 Significant variable according the strength

Variables has strong significant relationship	Variables has least significant relationship	Variables which have zero variance
Lack of maintenance	Improper use of material	Poor detailing
Vandalism	Soil movement impact	Faulty drawing
Poor waterproofing	Excessive moisture	Building size
Lack of supervision	Change of usage	Building types
Lack of cleaning	Poor construction	Building orientation
Misuse by user	Specification	Lack of knowledge
	Reaction of thermal agent	Poor ventilation
	Poor workmanship	Poor operation
	Wear and tear	
	Structural	
	Faulty design	
	Reaction of chemical agent	
	Change of climatic condition	
	Insect attack	
	Reaction of biological agent	

7.1.4 *Objective 4: To Develop a Building Deterioration Prediction Model (BDPM) of the Building Condition*

The prediction model produced will benefit organizations such as PWD and schools management. BDPM plays an important role in managing the assets such as critical decision making, controlling the maintenance cost, time efficiency, enhanced accuracy, resources, as well as also providing guidelines for best practices in asset management.

The logistic regression model was developed in this study to predict the condition of the building. SPSS software was used to analyse and produce the regression model. The Z value is than transformed using link function to obtain the probability of the event occurring as follow:

$$\mathrm{P}\big[\text{Poor Building Condition}\big]: = \frac{e^z}{1 + e^z} \text{ and the value is between 0 and 1}$$

Based on the result, only one model was produced to predict the probability of the building condition rating. There are 3 predictors that significantly improve the model as a predictor which is poor water proofing, lack of maintenance and lack of cleaning. The logistic regression equation with actual values of the regression coefficients produced is as follow:

$$\text{The Equation: } Z = -3.479 + 0.658(\text{PW}) + 0.755(\text{LOM}) + 0.445(\text{LOC})$$

As a conclusion, the variables significantly improve the model as predictors are poor waterproofing, lack of maintenance, and lack of cleaning. The model produced can predict the condition of the building but could not change the factors contributing towards the building defects. Those factors can be monitored to minimize or eliminate the defects. The true value of the prediction modeling can only be achieved when it is leveraged by operational users, for example asset managers, who applies the techniques to the information contained in databases and applications and uses the outcomes to make immediate operational decisions. Hence, it is very crucial to monitor the critical factors such as poor water proofing, lack of maintenance and lack of information for the designers and asset managers to solve maintenance issues within budget allocation. In order to improve the current maintenance management processes, especially involving human resources, such as building inspector with the right skills, BDPM's statistical analysis can be used easily.

7.2 Contribution to Knowledge

This thesis presented and aims to extend the body of knowledge in the BDPM of government buildings. The importance of the current condition of the building is always neglected by the user or client. The organization has to improve the existing or current maintenance system by implementing a strategic asset management method effectively. BCA is one of the tools used to enhance the maintenance management system for government building. Regression model produced as a tool to enhance BCA output used to forecast changes or deterioration of the condition of the building. Based on the study, this research makes a number of contribution to knowledge as follows:

(i) Benefit for the organization: The proposed model will benefit the organizations, such as PWD to monitor the asset under their responsibility. Using the model makes the process of condition assessment of the building fast. This will make better performance of the organization and the management
(ii) Improvement of the current maintenance management processes: The goal of this study was to improve the current procedure of inspection and condition assessment process for buildings. The current processess will be improved based on the findings in the study, for example the list of factors and defects affect the building condition.
(iii) Enhancement of current process of building assessment: The current process will be reviewed and the enhancement will be made on the updated data gathered in the study.
(iv) Better decision making in maintenance or replacement before failure: The prediction on the condition of the building will make better decision making in maintenance and save time.
(v) Efficient reduction in additional budget and saving in time, resources, operation budget and maintenance cost.

7.3 Recommendations for Future Research

This research aimed to produce a reliable model for building deterioration prediction model (BDPM) for school building, but hence, some aspect were beyond the scope of the project. Therefore, the research provides more opportunities to extend and continue to improve this research. Further study can be explored in the following fields:

(i) An in-depth study should be carried out to identify the type of defects that significantly influence the building condition. These variables are useful to generate a reliable BDPM.

(ii) The condition rating results from the condition assessment of building should be studied. The condition rating has been filtered for development of the model. The level of data collected and confidence of the data influences the model. It is recommended that the human resources skill and competency should be studied in order to ensure the reliability of data collected and improved outcomes of the model.

(iii) Similar study to be carried out with different scope of study. The study can focus on the prediction for cost of maintenance or rectification work according to the condition of the building.

(iv) Similar study with different scope of study, such as for other types of building with different design and shape such as an office building or health building should be carried out.

(v) A detailed study on the other tools such as systems and technology that can adopt and generate the BDPM integration with current government systems. For example, Geographical Information System (GIS) based reporting which is beneficial for identifying relationships in the data, that will facilitate the impact of the condition of the building and assist with the consequent decisions.

(vi) Detailed study of the current process of the BCA, which is possible to accommodate the level of confidence and increased accuracy of the prediction analysis.

Appendix A
Building Condition Assessment Checklist

S. Yacob et al., *Managing Building Deterioration*, Management in the Built Environment,
https://doi.org/10.1007/978-981-16-5860-0

JKR

BUILDING INFORMATION FORM

Asset Registration Code:	Building Category	1	2	3	4	Sheets:

A - BUILDING INSPECTION SERVICES FILL IN THE BLANK

1.1 Project Title / Work

1.2 Reference No.

1.3 Name of Inspector

1.4 Date of Inspection

B - BUILDING INFORMATION FILL IN THE BLANK

2.1 Name of Building

2.2 Owner

2.3 Address /Road/Lot

2.4 Block No: 2.5 Area 2.6 Age & Year Constructed

2.7 Main Material 2.8 Yearly Maintenance Cost

2.9 Building Structure 2.10 Operation Status

C - REFERENCE / COMPULSORY DOCUMENT FILL IN THE BLANK & TICK (√)/(X) IN RELATED BOXES

DESCRIPTION	TICK (√) / (X)	COMMENT	REFERENCES
3.1 Drawing approval from local Authority			
3.2 As Built Drawing			
3.3 Certificate / Warranty			
3.4 Maintenance Record			
3.5 Inspection Record			
3.6 Related Manual			
3.7 Act & Regulation			

NOTE - BUILDING CATEGORY

CODE	BUILDING CATEGORY	DESCRIPTION
1	Operation	Used for Department core service (example : Police Station, Hospital, Military, School)
2	Support	Used for function support to department (example : Library, Kem Latihan, Multi Storey Carpark, Administration Block)
3	Abandon	Abandoned (example : Building waiting to demolish)
4	Control	Building Govern by Government (example : Heritage Building)

INSPECTION CHECKLIST FOR CIVIL AND STRUCTURE WORK

project Title :	Block No :	Level:

PARTICULAR OF INSPECTION FILL IN THE BLANK OR CIRCLE THE ANSWERS

Inspected By:

Date : Day | Month | Year

Time : (Circle the time) AM / PM

Wheather : HOT | CLOUDY | RAIN | HEAVY RAIN

ID DPA No :

Space type :

INSPECTION FORM TICK (√) IN RELATED BOX

No	System		Subsystem		Component		Building Physical Condition					Maintenance Priority Assessment					Remarks
							VG(1)	G(2)	F(3)	C(4)	VC(5)	RP(5)	RH(4)	RE(3)	R(2)	N(1)	
1	01	Foundation	0101	Piling													
			0102	Pad foundation													
			0105	Ground Floor	001	Finishes											
2	02	Structure	0201	Upper Floor	001	Finishes											
			0202	Suspended Floor	001	Finishes											
			0203	Column	001	Finishes											
					002	Plaster											
					003	Paint											
			0204	Beam	001	Plaster											
					002	Paint											
			0103	External Staircase	001	Structure											
					002	Finishes											
					003	Tread											
					004	Riser											
					005	Balustrate											
					006	Handrail											
					007	Paint											
			0206	Ramp	001	Finishes											
3	03	*Exterior enclosure*	0301	External wall	001	Finishes											
					002	Plaster											
					003	Paint											
					004	Ornamen											
			0302	Parapet	001	Finishes											
					002	Plaster											
					003	Paint											

NO.	ELEMENTS / COMPONENTS	DEFECTS / DAMAGES	BCARS				
			Physical Condition [a]	Priority Assessment [b]	Matrix Analysis [c] (a x b)	Defect Sheet	Defect Identification Code
1							
2							
3							
4							
5							
6							
7							
8							
9							
10							
11							
12							
13							
14							
15							
16							
17							
18							
19							
20							
21							
22							
23							
24							
25							
Total Marks [d] (∑ of c)					0		
Total of Finding / Defects (d/e)					0		
Total Score (d/e)					**0.00**		
Overall Condition Rating							

Appendix B
Total Sampling for Population of School

S. Yacob et al., *Managing Building Deterioration*, Management in the Built Environment,
https://doi.org/10.1007/978-981-16-5860-0

Year	Building age	States											
		Selangor		N. Sembilan		Melaka		Johor		Kedah		Perak
		S	P	S	P	S	P	S	P	S	P	S
1995	20	5	14	0	1	0	1	3	6	2	6	5
1996	19	0	3	3	2	0	2	3	11	4	2	4
1997	18	6	17	0	0	0	1	1	2	1	3	4
1998	17	3	12	1	4	0	1	3	4	2	2	3
1999	16	9	5	2	3	1	3	2	7	4	6	3
2000	15	12	13	4	8	4	4	11	18	9	6	3
2001	14	13	24	2	3	1	2	15	27	5	8	7
2002	13	10	27	5	6	2	1	11	14	6	8	5
2003	12	17	19	1	1	0	1	10	5	5	1	3
2004	11	8	21	0	3	0	2	1	5	1	2	1
Total		83	155	18	31	8	18	60	99	39	44	38

States

Year	Building age	Perak	Pulau Pinang		Perlis		Kelantan		Terengganu		Pahang	
		P	S	P	S	P	S	P	S	P	S	P
1995	20	7	3	7	0	0	0	3	3	3	2	5
1996	19	3	2	3	2	1	0	1	1	3	0	3
1997	18	3	0	3	0	0	0	0	4	4	3	5
1998	17	2	1	2	0	1	0	0	4	1	3	4
1999	16	0	0	0	0	3	0	0	3	3	6	4
2000	15	3	3	3	2	0	0	0	3	4	10	5
2001	14	5	4	5	1	3	0	3	3	13	15	17
2002	13	9	7	9	0	1	0	2	3	4	11	15
2003	12	4	4	4	0	0	1	1	10	12	10	12

(continued)

(continued)

Year	Building age	States																			
		Selangor						N. Sembilan		Melaka				Johor				Kedah			Perak
		S		P				S	P	S		P		S		P		S		P	S
2004	11	2	2		2	C	0			0	0		3		4		4		1		
Total		38	26		38	5	9			1	10		37		51		64		71		
Overall total		974																			
Sample size		303																			

Appendix C
Assessment of RII for Factors, Defects and Condition Rating

S. Yacob et al., *Managing Building Deterioration*, Management in the Built Environment,
https://doi.org/10.1007/978-981-16-5860-0

BUILDING : SEKOLAH MENENGAH KEBANGSAAN KUALA NERANG, KEDAH (0124)

NO	ELEMENT / COMPONENT	DEFECTS	POSSIBLES FACTORS CONTRIBUTE TO DEFECT																													CONDITION RATING (a)	MAINTENANCE PRIORITY (b)	MATRIX ANALYSIS '(c) (a x b)
			TECHNICAL												ENVIRONMENT								HUMAN											
			Design	Detailing	Drawing	Specification	Structural	Waterproofing	Material	Maintenance	Construction	Building Size	Type of Building	Building Elevation	Insect Attack	Biological Agent	Climatic Condition	Ventilation	Thermal Agent	Moisture	Chemical Agent	Soil Impact	User	Vandalism	Wear & Tear	Usage	Workmanship	Lack of Supervision	Lack of Knowledge	Operation	Cleaning			
1	Partition - Timber	Termite Attack							5						5																	5	3	15
2	Break glass	Broken / Missing								5														4				5				5	5	25
3	Fire Extinguisher	Validity Certificate Expired								5																		5				3	5	15
4	Fire alarm Panel	Not Functioning								5																		5				4	5	22
5	Emergency Lamp	Not Functioning								5																		5				5	5	25
6	ELR	Validity Certificate Expired								5																		5				5	5	25
7	ELR	Validity Certificate Expired								5																		5				5	5	25
8	Ceiling	Dampness / Mold							3							3				3												4	3	11
9	Door	Broken / Missing							3															4								5	3	15
10	Window	Broken								3														5								3	3	8
11	Wall	Crack									2																4					4	3	13
12	Roofing	Leakage								4							4											3				4	3	12
13	Slab	Broken / Spalling																									4	4				3	5	15
14	Sanitary Pipe	Blockage								4																		4				4	3	12
15	Service pipe	Leakage								3																	4	3				3	3	7.5
16	Ceiling	Broken / Mold							4							5				5												3	3	10
17	Stage Lighting	Not Functioning								4																3		3				3	3	10
		w =	0	0	0	0	0	0	15	48	2	0	0	0	5	8	4	0	0	8	0	0	0	13	0	3	12	47	0	0	0	TOTAL MARKS [d] (Σ of c)		265
		A =	5	5	5	5	5	5	5	5	5	5	5	5	5	5	5	5	5	5	5	5	5	5	5	5	5	5	5	5	5	TOTAL OF FINDINGS [e]		17
		N =	17	17	17	17	17	17	17	17	17	17	17	17	17	17	17	17	17	17	17	17	17	17	17	17	17	17	17	17	17	TOTAL SCORE (d/e)		15.57
		n =							4	11	1				1	2	1			2				3		1	3	11				OVERALL RATINGS		C
KD01		RII =							0.75	0.87	0.40				1.00	0.80	0.80			0.80				0.87		0.60	0.80	0.85						
		Rank							5	2	7				1	4	4			4				2		6	4	3						

Appendix D
Semi-structured Interview Information Sheet

UNIVERSITY OF MALAYA

Information Sheet

Interviewer's Information

Research Title:	Development A Building Deterioration Prediction Model (BDPM) For Public Schools in Peninsular Malaysia
Researcher:	Syamilah bt Yacob (BHA 140001)
Faculty:	Faculty of Built Environment, University of Malaya

Brief of the project:

The study is conducted to explore the contribution factors to the building defects toward the condition of the school building. The relationship between factors contribute to the building defects and condition rating is studied to measure the significant factors toward building condition. At first stage data collected through building survey and content analysis for the archived documentation to identify the various factors as variables in this research. At the second stage, semi-structured interview to be conducted with participant of Public Work Department (PWD) involving professional at technical and management level. The information gathered through the discussion will be used to discuss the contribution factors towards building condition. Then, the information will be compared, merged and validated with a building survey results. The relationship between the factors towards building condition rating is expected to be established. As a result, for overall study, a prediction model will be developed and recommended.

S. Yacob et al., *Managing Building Deterioration*, Management in the Built Environment,
https://doi.org/10.1007/978-981-16-5860-0

Aims of study:

The aim of this research is to study on the condition of the building, including the factors contribute to the building defects. The contributing factors to building defects is to be identified towards the building condition rating.

The objectives of the research are:

(a) To identify the factors contributing to the building defects;
(b) To investigate the condition of the school buildings through building condition assessment;
(c) To establish significant relationship between the factors contributing to the building defects and building condition;
(d) To develop a building deterioration prediction model (BDPM) on the building condition.

Requirements:

(a) Participant must be from the Public Work Department (PWD) Malaysia;
(b) Participate in this discussion must be professional level and as appointed building inspector for Building Condition Assessment (BCA);
(c) Have working experience of more than 3 years in conducting BCA for government buildings;
(d) The participant will be Building Surveyor and Engineer at technical and management level; and
(e) Participant will be informed for the session of discussions and bound with consent form for participation.

Confidentiality:

(a) All the discussion with participant are confidential;
(b) The identity of the participant will be abstracted in the thesis; and
(c) All data collected is gathered in this participation only be assessed by the researcher and her supervisors.

UNIVERSITY OF MALAYA

CONSENT FORM

Research Title:	Developing a Prediction Model for Building Deterioration: The Case of School Building in Malaysia
Researcher:	Syamilah bt Yacob (BHA 140001)
Faculty:	Faculty of Built Environment, University of Malaya

I have been given the information about the research and I have been informed and understand the purposes of the study and my role as a participant. I understand that my participation in this research is voluntary and I have had right to ask the researcher any questions. I am also free to refuse to participate in the research at any time without prejudices.

I understand that the interview will be audio recorded/documented to facilitate the collection of information with the understanding that all information which I provide will be held in confidence. I understand that the data collected from my participation will be used for a PhD thesis and will also be used in summary form for journal publication, and I consent for it to be used in that manner.

By signing below I am indicating my consent to participate in the research

Name: ____________________________________

Job Title: ____________________________________

Signature: ____________________________________

Date: ____________________________________

Interview Questions

Question 1

Section A

Based on the research, a total of 40 factors was founded in the literature which caused the building defect. The factors have been arranged accordingly to group base similarities to be 29 number of factors. From the analysis, it found that from the 29 number of factors analysed, only 21 factors were analysed to identify the significant relationship between factors and building condition rating. There are 6 factors which have the most significant correlation between factors and building condition. How do you rate the following factors would affect the overall condition of the building? Give your comment for your answer.

Lack of maintenance
Highly disagree 0 0 0 0 0 Highly agree
Vandalism
Highly disagree 0 0 0 0 0 Highly agree
Poor waterproofing
Highly disagree 0 0 0 0 0 Highly agree
Lack of supervision
Highly disagree 0 0 0 0 0 Highly agree
Lack of cleaning
Highly disagree 0 0 0 0 0 Highly agree
Misuse by user
Highly disagree 0 0 0 0 0 Highly agree

Legend:
0 0 0 0 0
1 2 3 4 5
1—Highly disagree
2—Disagree
3—Neither agree/disagree
4—Agree
5—Highly agree

Section B

Other 15 number of factors analysed, it is found that those factors have at least (least) significant relationship between factors contribute to building defects and building condition rating. How do you rate the following factors would affect the overall condition of the building? Give your comment for your answer.

Improper use of material
Highly disagree 0 0 0 0 0 Highly agree
Insect attack
Highly disagree 0 0 0 0 0 Highly agree
Faulty design
Highly disagree 0 0 0 0 0 Highly agree
Wear and tear
Highly disagree 0 0 0 0 0 Highly agree
Poor workmanship
Highly disagree 0 0 0 0 0 Highly agree
Not complying with specification
Highly disagree 0 0 0 0 0 Highly agree
Reaction thermal agent
Highly disagree 0 0 0 0 0 Highly agree
Excessive moisture
Highly disagree 0 0 0 0 0 Highly agree

Legend:
0 0 0 0 0
1 2 3 4 5
1—Highly disagree
2—Disagree
3—Neither agree/disagree
4—Agree
5—Highly agree

Poor construction
Highly disagree 0 0 0 0 0 Highly agree
Change of uses
Highly disagree 0 0 0 0 0 Highly agree
Soil movement impact
Highly disagree 0 0 0 0 0 Highly agree
Structural
Highly disagree 0 0 0 0 0 Highly agree
Change of climatic condition
Highly disagree 0 0 0 0 0 Highly agree
Reaction of chemical agent
Highly disagree 0 0 0 0 0 Highly agree
Reaction of biological agent
Highly disagree 0 0 0 0 0 Highly agree

Legend:
0 0 0 0 0
1 2 3 4 5
1—Highly disagree
2—Disagree
3—Neither agree/disagree
4—Agree
5—Highly agree

Section C

From the analysis of the factors such as detailing, drawing, specification, structural, building size, building types, building elevation, lack of knowledge, and operation has zero variance and is removed from the scale. How do you rate the following factors would affect the overall condition of the building? Give your comment for your answer.

Poor detailing
Highly disagree 0 0 0 0 0 Highly agree
Faulty drawing
Highly disagree 0 0 0 0 0 Highly agree
Building size
Highly disagree 0 0 0 0 0 Highly agree
Types of building
Highly disagree 0 0 0 0 0 Highly agree
Building orientation
Highly disagree 0 0 0 0 0 Highly agree
Ventilation
Highly disagree 0 0 0 0 0 Highly agree
Lack of knowledge
Highly disagree 0 0 0 0 0 Highly agree
Operation
Highly disagree 0 0 0 0 0 Highly agree

Legend:
0 0 0 0 0
1 2 3 4 5
1—Highly disagree
2—Disagree
3—Neither agree/disagree
4—Agree
5—Highly agree

Question 2

To isolate and identify the most likely causes based on the background of the study, the theoretical framework of the actual problem as possible independent, mediating and dependent variables was developed based on literature and secondary data (BCA Guideline, report of inspection). (The researcher will explain the theoretical framework of the study as attached).

Are you agree with the variables identified would lead to various types of building defects which can affect the building condition rating? Give your comment (allow for the suggestion).

(a) How these factors affect the building defects towards building condition?
(b) Do you agree the most common defects shown are usually found in government buildings?
(c) In your opinion, is the factors related each other to contribute to the building defects?

Question 3

To assess the condition of the building, the guidelines and systems the Building Condition Inspection of Existing Building (2013) were adopted in the study. BCA rating commonly uses a weighting factor to represent the severity of each defect.

(a) What is your opinion about the BCA?
(b) How do you rate the extent of the defects in your inspection?
(c) How do you do to get the accuracy of building condition rating?

The End & Thank you

Theoretical Framework

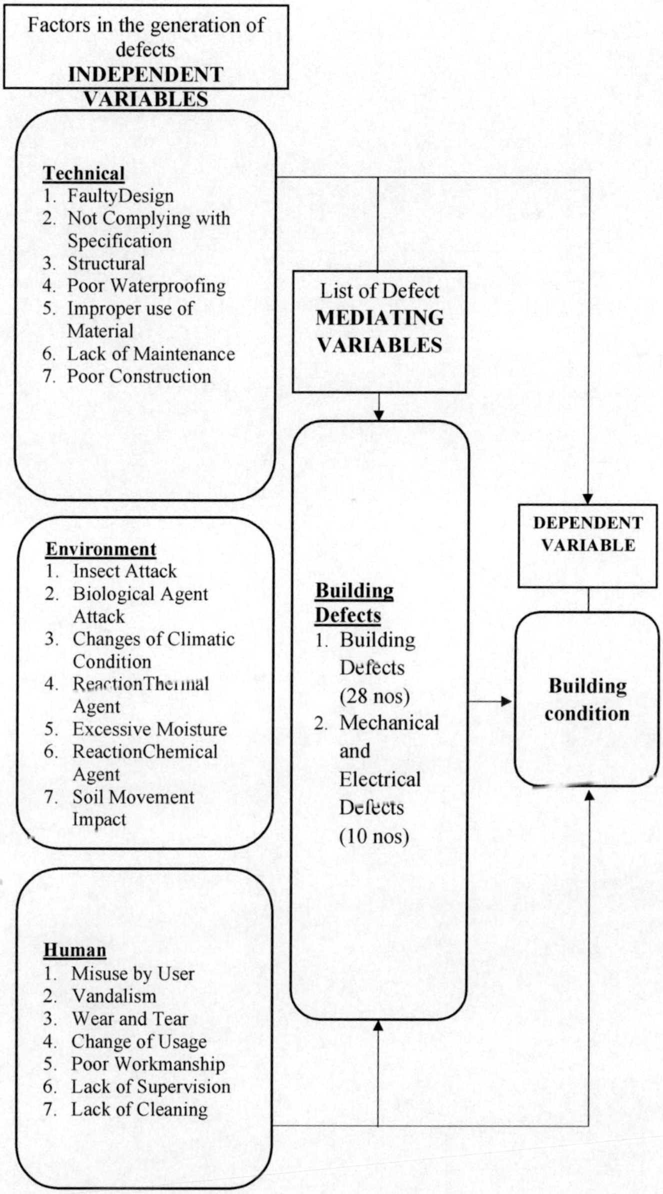

Appendix E
Reliability Analysis

Case processing summary

		N	%
Cases	Valid	303	100.0
	Excluded[a]	0	0
	Total	303	100.0

[a]Listwise deletion based on all variables in the procedure

Reliability statistics

Cronbach's Alpha	Cronbach's Alpha based on standardized items	N of items
0.713	0.712	21

Item statistics

	Mean	Std. deviation	N
Faulty design	1.6271	1.15210	303
Not complying with specification	1.4455	0.93294	303
Structural	1.4587	0.93715	303
Poor waterproofing	2.7228	1.60578	303
Improper use of material	3.4422	0.71581	303
Lack of maintenance	4.5710	0.70058	303
Poor construction	3.1914	1.09317	303

(continued)

S. Yacob et al., *Managing Building Deterioration*, Management in the Built Environment,
https://doi.org/10.1007/978-981-16-5860-0

(continued)

	Mean	Std. deviation	N
Insect attack	2.3861	1.70878	303
Reaction of biological agent	2.5941	1.14953	303
Change of climatic condition	2.7162	1.11817	303
Reaction of thermal agent	1.4587	0.94770	303
Excessive moisture	3.0462	1.08474	303
Reaction of chemical agent	1.7492	1.11126	303
Soil impact movement	2.0594	1.18340	303
Misuse by user	2.6799	1.25538	303
Vandalism	3.8680	1.16619	303
Wear and tear	3.2013	1.21894	303
Change of uses	1.6139	1.03550	303
Poor workmanship	3.3696	1.03674	303
Lack of supervision	4.5809	0.88357	303
Lack of cleaning	1.9175	1.20008	303

Item-total statistics

	Scale mean if item deleted	Scale variance if item deleted	Corrected item-total correlation	Squared multiple correlation	Cronbach's Alpha if Item deleted
Faulty design	54.0726	73.399	0.454	0.358	0.687
Not complying with specification	54.2541	74.117	0.541	0.641	0.684
Structural	54.2409	74.965	0.483	0.467	0.688
Poor waterproofing	52.9769	76.122	0.178	0.146	0.717
Improper use of material	52.2574	81.834	0.103	0.327	0.714
Lack of maintenance	51.1287	85.404	−0.171	0.303	0.728
Poor construction	52.5083	78.251	0.219	0.374	0.708
Insect attack	53.3135	76.017	0.159	0.342	0.721
Reaction of biological agent	53.1056	78.797	0.175	0.427	0.712
Change of climatic condition	52.9835	78.135	0.217	0.142	0.708

(continued)

(continued)

	Scale mean if item deleted	Scale variance if item deleted	Corrected item-total correlation	Squared multiple correlation	Cronbach's Alpha if Item deleted
Reaction of thermal agent	54.2409	73.621	0.563	0.612	0.682
Excessive moisture	52.6535	79.466	0.157	0.432	0.713
Reaction of chemical agent	53.9505	75.729	0.347	0.288	0.697
Soil impact movement	53.6403	76.920	0.258	0.196	0.704
Misuse by user	53.0198	72.847	0.432	0.383	0.688
Vandalism	51.8317	78.101	0.205	0.149	0.709
Wear and tear	52.4983	71.344	0.527	0.458	0.678
Change of uses	54.0858	73.337	0.523	0.526	0.683
Poor workmanship	52.3300	78.513	0.223	0.306	0.707
Lack of supervision	51.1188	84.860	−0.120	0.261	0.729
Lack of cleaning	53.7822	75.005	0.348	0.288	0.696

Summary item statistics

	Mean	Min	Max	Range	Max/min	Variance	N of items
Item means	2.652	1.446	4.581	3.135	3.169	0.950	21
Item variances	1.278	0.491	2.920	2.429	5.949	0.330	21
Inter-item Covariances	0.135	−0.402	0.773	1.176	– 1.922	0.039	21
Inter-item Correlations	0.105	−0.311	0.674	0.985	– 2.168	0.032	21

Scale statistics

Mean	Variance	Std. deviation	N of items
55.6997	83.681	9.14773	21

Appendix F
Logistic Regression Analysis

Case processing summary

Unweighted cases[a]		N	Percentage
Selected cases	Included in analysis	303	100.0
	Missing cases	0	0
	Total	303	100.0
Unselected cases		0	0
Total		303	100.0

[a]If weight is in effect, see classification table for the total number of cases

Dependent variable encoding

Original value	Internal value
1.00	0
2.00	1

S. Yacob et al., *Managing Building Deterioration*, Management in the Built Environment,
https://doi.org/10.1007/978-981-16-5860-0

Block 0: Beginning Block

Classification table[a,b]

Observed			Predicted		
			NEWCON		Percentage correct
			1.00	2.00	
Step 0	NEWCON	1.00	0	16	0
		2.00	0	287	100.0
	Overall percentage				94.7

[a]Constant is included in the model
[b]The cut value is 0.500

Variables in the equation

		B	S.E	Wald	df	Sig	Exp(B)
Step 0	Constant	2.887	0.257	126.305	1	0.000	17.937

Variables not in the equation

			Score	df	Sig
Step 0	Variables	Poor waterproofing	5.446	1	0.020
		Lack of maintenance	13.856	1	0.000
		Vandalism	0.173	1	0.677
		Lack of supervision	8.985	1	0.003
		Misuse by user	0.348	1	0.555
		Lack of cleaning	5.891	1	0.015
	Overall statistics		25.555	6	0.000

Block 1: Method = Forward Stepwise (Likelihood Ratio)

Omnibus tests of model coefficients

		Chi-square	df	Sig
Step 1	Step	9.760	1	0.002
	Block	9.760	1	0.002
	Model	9.760	1	0.002
Step 2	Step	6.524	1	0.011
	Block	16.284	2	0.000
	Model	16.284	2	0.000
Step 3	Step	4.109	1	0.043

(continued)

(continued)

		Chi-square	df	Sig
	Block	20.394	3	0.000
	Model	20.394	3	0.000

Model summary

Step	−2 log likelihood	Cox & Snell R Square	Nagelkerke R Square
1	115.496[a]	0.032	0.094
2	108.972[b]	0.052	0.155
3	104.863[b]	0.065	0.192

Hosmer and lemeshow test

Chi-square	df	Sig.
0.005	1	0.945
2.697	6	0.846
6.624	8	0.578

Contingency table for hosmer and lemeshow test

		NEWCON = 1.00		NEWCON = 2.00		Total
		Observed	Expected	Observed	Expected	
Step 1	1	4	4.042	14	13.958	18
	2	6	5.873	79	79.127	85
	3	6	6.086	194	193.914	200
Step 2	1	4	3.772	11	11.228	15
	2	4	4.107	29	28.893	33
	3	0	0.967	12	11.033	12
	4	5	4.247	75	75.753	80
	5	1	1.298	29	28.702	30
	6	1	0.848	43	43.152	44
	7	1	0.384	28	28.616	29
	8	0	0.378	60	59.622	60
Step 3	1	6	6.530	23	22.470	29
	2	3	2.175	17	17.825	20
	3	2	2.163	26	25.837	28
	4	1	0.671	12	12.329	13

(continued)

(continued)

		NEWCON = 1.00		NEWCON = 2.00		Total
		Observed	Expected	Observed	Expected	
	5	2	2.587	63	62.413	65
	6	0	0.870	31	30.130	31
	7	2	0.505	30	31.495	32
	8	0	0.302	31	30.698	31
	9	0	0.078	14	13.922	14
	10	0	0.119	40	39.881	40

Classification table[a]

	Observed		Predicted		
			NEWCON		Percentage correct
			1.00	2.00	
Step 1	NEWCON	1.00	0	16	0.0
		2.00	0	287	100.0
	Overall percentage				94.7
Step 2	NEWCON	1.00	0	16	0.0
		2.00	0	287	100.0
	Overall percentage				94.7
Step 3	NEWCON	1.00	0	16	0.0
		2.00	0	287	100.0
	Overall percentage				94.7

[a]The cut value is 0.500

Variables in the equation

		B	S.E.	Wald	df	Sig.	Exp(B)	95% C.I. for EXP(B)	
								Lower	Upper
Step 1[a]	Lack of maintenance	0.861	0.257	11.230	1	0.001	2.365	1.429	3.912
	Constant	−0.842	1.078	0.611	1	0.435	0.431		
Step 2[b]	Poor waterproofing	0.545	0.245	4.952	1	0.026	1.725	1.067	2.787
	Lack of maintenance	0.930	0.274	11.540	1	0.001	2.535	1.482	4.336
	Constant	−2.315	1.323	3.061	1	0.080	0.099		
Step 3[c]	Poor waterproofing	0.658	0.262	6.294	1	0.012	1.932	1.155	3.231

(continued)

(continued)

		B	S.E.	Wald	df	Sig.	Exp(B)	95% C.I. for EXP(B)	
								Lower	Upper
	Lack of maintenance	0.755	0.288	6.893	1	0.009	2.129	1.211	3.741
	Lack of cleaning	0.445	0.213	4.388	1	0.036	1.561	1.029	2.368
	Constant	−3.479	1.484	5.496	1	0.019	0.031		

[a]Variable(s) entered on step 1: Lack of Maintenance
[b]Variable(s) entered on step 2: Poor Waterproofing
[c]Variable(s) entered on step 3: Lack of Cleaning

Correlation matrix

		Constant	Lack of maintenance	Poor waterproofing	Lack of maintenance	Lack of cleaning
Step 1	Constant	1.000	−0.970			
	AMAINTENANCE	−0.970	1.000			
Step 2	Constant	1.000		−0.553	−0.924	
	Poor waterproofing	−0.553		1.000	0.260	
	AMAINTENANCE	−0.924		0.260	1.000	
Step 3	Constant	1.000		−0.609	−0.734	−0.391
	Poor waterproofing	−0.609		1.000	0.215	0.238
	AMAINTENANCE	−0.734		0.215	1.000	−0.238
	Lack of cleaning	−0.391		0.238	−0.238	1.000

Model if term removed

Variable		Model log likelihood	Change in −2 log likelihood	df	Sig. of the change
Step 1	Lack of maintenance	−62.628	9.760	1	0.002
Step 2	Poor waterproofing	−57.748	6.524	1	0.011
	Lack of maintenance	−59.626	10.280	1	0.001
Step 3	Poor waterproofing	−56.777	8.691	1	0.003
	Lack of maintenance	−55.626	6.389	1	0.011
	Lack of cleaning	−54.486	4.109	1	0.043

Variables not in the equation

			Score	df	Sig
Step 1	Variables	Poor waterproofing	5.757	1	0.016
		Vandalism	0.000	1	1.000
		Lack of supervision	2.081	1	0.149
		Misuse by user	0.998	1	0.318
		Lack of cleaning	2.092	1	0.148
	Overall statistics		13.425	5	0.020
Step 2	Variables	Vandalism	0.267	1	0.606
		Lack of supervision	1.212	1	0.271
		Misuse by user	0.509	1	0.476
		Lack of cleaning	4.618	1	0.032
	Overall statistics		7.678	4	0.104
Step 3	Variables	Vandalism	0.091	1	0.763
		Lack of supervision	0.140	1	0.708
		Misuse by user	2.078	1	0.149
	Overall statistics		2.964	3	0.397

```
          Step number: 1

          Observed Groups and Predicted Probabilities

     200 +                                                                  2  +
         I                                                                  2  I
         I                                                                  2  I
F        I                                                                  2  I
R    150 +                                                                  2  +
E        I                                                                  2  I
Q        I                                                                  2  I
U        I                                                                  2  I
E    100 +                                                                  2  +
N        I                                                               2  2  I
C        I                                                               2  2  I
Y        I                                                               2  2  I
      50 +                                                               2  2  +
         I                                                               2  2  I
         I                                                               2  2  I
         I                                        2                 2    2  2  I
Predicted ---------+---------+---------+---------+---------+---------+---------+---------+---------+----------
  Prob:   0        .1        .2        .3        .4        .5        .6        .7        .8        .9         1
  Group:  111111111111111111111111111111111111111111111111111222222222222222222222222222222222222222222222222222

          Predicted Probability is of Membership for 2.00
          The Cut Value is .50
          Symbols: 1 - 1.00
                   2 - 2.00
          Each Symbol Represents 12.5 Cases.
```

```
          Step number: 2

          Observed Groups and Predicted Probabilities

      80 +                                                                                              2     +
         I                                                                                              2     I
         I                                                                                              2   2 I
F        I                                                                                              2   2 I
R     60 +                                                                                              2   22+
E        I                                                                                              2   22I
Q        I                                                                                              2   22I
U        I                                                                                              2   22I
E     40 +                                                                                              2   22+
N        I                                                                                       2      2   22I
C        I                                                                                       2      2   22I
Y        I                                                                                       2      22  22I
      20 +                                                                                       2      22  22+
         I                                                                                       2      22  22I
         I                                                                                       2    2 22  22I
         I                                                                            2  2     2   1    2 122222I
Predicted ---------+---------+---------+---------+---------+---------+---------+---------+---------+----------
  Prob:   0        .1        .2        .3        .4        .5        .6        .7        .8        .9         1
  Group:  1111111111111111111111111111111111111111111111111122222222222222222222222222222222222222222222222222

          Predicted Probability is of Membership for 2.00
          The Cut Value is .50
          Symbols: 1 - 1.00
                   2 - 2.00
          Each Symbol Represents 5 Cases.
```


```
          Step number: 2

          Observed Groups and Predicted Probabilities

      80 +                                                                                              2     +
         I                                                                                              2     I
         I                                                                                              2   2 I
F        I                                                                                              2   2 I
R     60 +                                                                                              2   22+
E        I                                                                                              2   22I
Q        I                                                                                              2   22I
U        I                                                                                              2   22I
E     40 +                                                                                              2   22+
N        I                                                                                       2      2   22I
C        I                                                                                       2      2   22I
Y        I                                                                                       2      22  22I
      20 +                                                                                       2      22  22+
         I                                                                                       2      22  22I
         I                                                                                       2    2 22  22I
         I                                                                            2  2     2   1    2 122222I
Predicted ---------+---------+---------+---------+---------+---------+---------+---------+---------+----------
  Prob:   0        .1        .2        .3        .4        .5        .6        .7        .8        .9         1
  Group:  1111111111111111111111111111111111111111111111111122222222222222222222222222222222222222222222222222

          Predicted Probability is of Membership for 2.00
          The Cut Value is .50
          Symbols: 1 - 1.00
                   2 - 2.00
          Each Symbol Represents 5 Cases.
```